Engaging Young Students in

Mathematics
through Competitions

World Perspectives and Practices

Volume I
Competition-ready Mathematics

Problem Solving in Mathematics and Beyond

Print ISSN: 2591-7234
Online ISSN: 2591-7242

Series Editor: Dr. Alfred S. Posamentier
Distinguished Lecturer
New York City College of Technology - City University of New York

There are countless applications that would be considered problem solving in mathematics and beyond. One could even argue that most of mathematics in one way or another involves solving problems. However, this series is intended to be of interest to the general audience with the sole purpose of demonstrating the power and beauty of mathematics through clever problem-solving experiences.

Each of the books will be aimed at the general audience, which implies that the writing level will be such that it will not engulfed in technical language — rather the language will be simple everyday language so that the focus can remain on the content and not be distracted by unnecessarily sophiscated language. Again, the primary purpose of this series is to approach the topic of mathematics problem-solving in a most appealing and attractive way in order to win more of the general public to appreciate his most important subject rather than to fear it. At the same time we expect that professionals in the scientific community will also find these books attractive, as they will provide many entertaining surprises for the unsuspecting reader.

Published

Vol. 13 *Engaging Young Students in Mathematics through Competitions —*
World Perspectives and Practices: Volume I — Competition-ready
Mathematics; Entertaining and Informative Problems from the
WFNMC8 Congress in Semriach/Austria 2018
edited by Robert Geretschläger

Vol. 12 *The Psychology of Problem Solving: The Background to Successful*
Mathematics Thinking
by Alfred S. Posamentier, Gary Kose, Danielle Sauro Virgadamo and
Kathleen Keefe-Cooperman

Vol. 11 *Solving Problems in Our Spatial World*
by Guenter Maresch and Alfred S. Posamentier

For the complete list of volumes in this series, please visit www.worldscientific.com/series/psmb

**Problem Solving in
Mathematics and Beyond** Volume **13**

Engaging Young Students in
Mathematics
through Competitions
World Perspectives and Practices

Volume I
Competition-ready Mathematics

Editor

Robert Geretschläger
BRG Kepler, Austria

World Scientific

NEW JERSEY · LONDON · SINGAPORE · BEIJING · SHANGHAI · HONG KONG · TAIPEI · CHENNAI · TOKYO

Published by

World Scientific Publishing Co. Pte. Ltd.

5 Toh Tuck Link, Singapore 596224

USA office: 27 Warren Street, Suite 401-402, Hackensack, NJ 07601

UK office: 57 Shelton Street, Covent Garden, London WC2H 9HE

Library of Congress Control Number: 2019030293

British Library Cataloguing-in-Publication Data
A catalogue record for this book is available from the British Library.

Problem Solving in Mathematics and Beyond — Vol. 13
ENGAGING YOUNG STUDENTS IN MATHEMATICS THROUGH
COMPETITIONS — WORLD PERSPECTIVES AND PRACTICES
Volume I — Competition-ready Mathematics; Entertaining and Informative Problems from
the WFNMC8 Congress in Semriach/Austria 2018

ISBN 978-981-120-582-8
ISBN 978-981-120-723-5 (pbk)

For any available supplementary material, please visit
https://www.worldscientific.com/worldscibooks/10.1142/11430#t=suppl

Desk Editors: V. Vishnu Mohan/Tan Rok Ting

Typeset by Stallion Press
Email: enquiries@stallionpress.com

Printed in Singapore

Foreword

Hello, Dear Reader!

What follows is a book, full of beautiful mathematics and ideas of how to convey this beauty to our students. Let me address in this brief introduction the issue that concerns me greatly, the ethical foundation of the Profession.

We, Mathematicians, occupy a special place in society. We are allowed to remain nerdy children as long as we inhabit our Ivory Tower and do not meddle in the affairs of our land and the world. In trying to shield me from problems with the regime in the Soviet Russia, my parents taught me that "For honest and talented people there are arts and sciences, and for the rest there is politics." In 1968, I was awakened from this parents' dream by Russian tanks rolling down streets of Prague. My 20-year long work on the book *The Scholar and the State* (Birkhäuser, 2015) further alerted me to the ethical problems of the world. I realized that there is no good science or good art, unless they are built on the foundation of high moral principles. I believe that Mathematicians are human beings, and as such ought to be citizens of their lands and citizens of the world. We cannot afford putting blinders on and seeing nothing outside of Mathematics.

L.E.J. Brouwer (1881–1966)

For those who believe that Mathematicians are entitled to live in the Ivory Tower and not contribute to the ethical and political discourse of the world, let me share the words of the greatest Dutch Mathematician of all time, Luitzen Egbertus Jan Brouwer:

"It is my opinion that the tiniest moral matter is more important than all of science, and that one can only maintain the moral quality of the world by standing up to any immoral project."

Millions of people prefer not to create waves and hide behind an inexcusable excuse "What can I do alone?" For the silent majority, I wish to quote Albert Einstein:

"The world is a dangerous place to live; not because of the people who are evil, but because of the people who don't do anything about it."

Albert Einstein (1879–1955)

Einstein's words explain precisely why Grisha Perelman left Mathematics. And there are others, less prominent colleagues, whose names may be not widely known. The 1986 Nobel Peace Prize Laureate and the Holocaust Survivor Elie Wiesel comes to mind with his call for action:

"There may be times when we are powerless to prevent injustice, but there must never be a time when we fail to protest."

In recent years, Schiller's famous words from Beethoven's last symphony have come to symbolize a protest against tyranny, as a hymn to the unity

Elie Wiesel (1928–2016)

of peoples based on actions for freedom, human rights, and mutual respect of the nations. Let us leave the Ivory Tower, join humanitarian activism, and sing along,

Alle Menschen werden Brüder.
Don't let tyrants rule the world!

Alexander Soifer
USA

About the Authors

 **Kiril Bankov** prepares future mathematics teachers as a Professor of Mathematics Education at the University of Sofia and the Bulgarian Academy of Sciences in Bulgaria. He graduated and received his Ph.D. in mathematics at the same University. Prof. Bankov has been involved in mathematics competitions in Bulgaria for more than 20 years as an author of contest problems and as a member of various juries. He has written many articles, made presentations, and is a co-author of books on mathematics competitions, problem solving, work with mathematically gifted students, etc. Two of his papers in *Mathematical Gazette* received awards as Article of the Year for 1995 and 1999. Prof. Bankov has a great deal of experience in international large-scale studies in mathematics education. For more than a decade he has been a member of the International Expert Committee (SMIRC) for the TIMSS study. He has also worked as a mathematics coordinator for the International Study Center of the Teacher Education and Development Study-Mathematics (TEDS-M) at Michigan State University (USA). In 2014 he was Chairperson of the European Baccalaureate Examining Board. Kiril Bankov was the Secretary of World Federation of National Mathematics Competitions (WFNMC) from 2008 till 2012. In 2012, he was elected as the Senior Vice-President of WFNMC and in July 2018 he became the President of the Federation.

Krzysztof Ciesielski works at the Mathematics Institute of the Jagiellonian University in Kraków, Poland. In 1999–2008, he was a Vice-Head of the Institute responsible for teaching students. His mathematical specialty is in topological dynamical systems. He works actively in the popularization of mathematics. He is author and co-author (mainly with *Zdzisław Pogoda*) of several books populariz-ing mathematics. Some of these have been awarded prestigious prizes in Poland; in particular, their recent book *Matematyczna bombonierka* ("Mathematical chocolate box") was recognized as the best book popularizing science by a Polish author in 2015/16 and awarded a Golden Rose Prize. In 2009–2016, he was a member of the Raising Public Awareness Committee of the European Mathematical Society, in 1999–2012, he was an Associate Editor of the *EMS Newsletter*. He is currently a Correspondent for *The Mathematical Intelligencer*, a Vice-Chair of the Editorial Board of the Polish monthly *Delta*, and the Editor-in-Chief of the journal *Wiadomości Matematyczne* of the Polish Mathematical Society. Since 1980 he has worked actively in the Kraków Regional Committee of the Polish Mathematical Olympiad, being its Head since 2008. Since 2002 he has been also a Vice-Head of Kraków Branch of Kangaroo Mathematical Competition. Since 2018 he has been a Vice-President of the WFNMC.

Hidetoshi Fukagawa works part time as a high school teacher of mathematics at Meiwa High School in Nagoya, Japan. He has spent many years studying a largely forgotten area of traditional mathematics, which includes a wealth of material for competitions and work with gifted students. He has published the books, *Traditional Japanese Mathematics: SANGAKU* in 1989 from C.B.R.C in Canada, *Traditional Japanese Mathematics Prob-lems of the 18th and 19th Centuries* in 2002 from SCT Publishing of Singapore, and *Sacred Mathematics: SANGAKU* in 2004 from Princeton University Press.

Robert Geretschläger teaches mathematics and descriptive geometry at BRG Kepler in Graz, Austria. He has been involved with various mathematics competitions in various capacities for decades, for instance as a trainer for Olympiads, as organizer for such things as the Austrian Kangaroo competition or as deputy leader and leader of the Austrian IMO team since 2000, and is the current Senior Vice-President of the WFNMC. He has also been involved in curriculum development, teacher training, textbook writing, and more. Despite keeping so busy, he is still usually able to keep a smile on his face.

Romualdas Kašuba teaches mathematics, communications skills and ethics at Vilnius University, Lithuania. He obtained his Ph.D. at the University of Greifswald in Germany. He has a long history with various Mathematical Olympiads: being a jury member of the Lithuanian MO since 1979, Deputy Leader of the Lithuanian IMO team from 1996 to 2017 and leader of the Lithuanian team at the Baltic Way team contest since 1995. Since 1996 he has been responsible for the Lithuanian team contest. In 1999, he initiated the Lithuanian Olympiad for youngsters and has been involved in the Lithuanian Kangaroo organization from the same year, as well as being leader of the Lithuanian MEMO team since 2009 and leader and deputy leader of the Lithuanian EGMO team since 2015. He is organizer, proposer and adapter of problems for many regional team contests in Lithuania. Besides being the author of several books in Lithuanian, he has authored several booklets in English (*What to Do When You Do Not Know What to Do, Parts I and II*, as well as *Once Upon a Time I Saw the Puzzle, Parts I, II and III*). Furthermore, he rewrote and expanded a book with a similar title in Russian, which was published in Moscow in 2012 (2nd edition 2014). From 2008 to 2016 he represented Lithuania at ICMI, and from 2010 to 2014 he was the Board member of MCG (the International Group for Mathematical Creativity and Giftedness) and is a member of the Editorial Board of the MCG Newsletter.

Marcin E. Kuczma teaches advanced calculus and various branches of analysis at the University of Warsaw. His main research areas are real variable functions and measure theory. He has been engaged in math competitions activities for decades, being strongly active at the Polish MO and the Austro-Polish mathematics competition. He has worked at eight IMOs as a member of the Problem Selection Committee and at twelve IMOs as coordinator. He was also Chair of the Jury at the Baltic Way Competition in 1998. He has been involved with the WFNMC since its beginning and was awarded the Hilbert Prize in 1992. Furthermore, he was chair of the Problems Section at the 3rd WFNMC Congress. He is the author of several books and many publications in the problem columns of several journals.

Edmundas Mazėtis received his Ph.D. from the University of Belarus in 1993. Since 1998, he has been an associate professor at the Lithuanian University of Educational Sciences, and from 2014 at Vilnius University. His main areas of scientific activity are in Differential Geometry and the Didactics of Mathematics. He is an organizer of the Lithuanian Regional Mathematical Olympiads, Head of the Lithuanian Young Mathematicians School, and a member of a group preparing Lithuanian pupils for international Olympiads.

Contents

Foreword   v

About the Authors   ix

Introduction   xv

**Part 1.   Some Paths Leading from Interesting Mathematics
to the Development of Potential Competition
Problems   1**

Chapter 1.1   Some Standard-Like Problems and
                    Non-standard Solutions   3
  Krzysztof Ciesielski

Chapter 1.2   Balls and Polyhedra   17
  Robert Geretschläger

Chapter 1.3   Hunting of Lions: Inversion May Help   47
  Kiril Bankov

Chapter 1.4   Sangaku: Traditional Japanese Mathematics   61
  Hidetoshi Fukagawa

Chapter 1.5 Can We Pose Problems That are Attractive,
            Yet Accessible to Many?   87

Edmundas Mazėtis and Romualdas Kašuba

Chapter 1.6 A Functional Equation Arising from
            Compatibility of Means   109

Marcin E. Kuczma

Chapter 1.7 Open Problems as Generalizations of Tasks
            from Mathematics Competitions   117

Kiril Bankov

**Part 2.   Some Favorite Puzzles and Problems Presented
           by Participants**   **125**

Chapter 2.1 Introduction, Problems and Solutions   127

Index   175

Introduction

The week of July 18–24, 2018 was an exciting time for the little village of Semriach, in the mountains near Graz in Austria. Although most of the villagers were most likely not aware of anything unusual transpiring, during the course of this week, Semriach played host to over 60 mathematicians, educators and problem creators from all over the world; participants in the eighth congress of the World Federation of National Mathematics Competitions (or WFNMC, for short).

The WFNMC is an international group, whose aim is to provide a common arena for all those concerned with organizing and developing problems for mathematics competitions, as well as related activities intended to stimulate the learning of mathematics. The organization has been active since 1984. Every other year, there is a meeting somewhere in the world, where interested parties are invited to actively participate in the advancement of the Federation's goals, be it by holding talks, taking part in discussions or joining in problem development sessions. If the year happens to be divisible by 4, the meeting is held in conjunction with that year's ICME (International Congress on Mathematical Education), as the WFNMC is an Affiliated Study Group of the ICMI (International Commission on Mathematical Instruction), the worldwide organization devoted to research and development in mathematical education at all levels, who also happen to organize the ICME meetings. In the intermediate even years, the WFNMC holds its own separate congresses, and the meeting in Austria was the latest in this series.

The material presented in the present two volumes is typical of the breadth of topic matter discussed at these meetings. Much of it involves fascinating mathematical ideas, as they are used to prepare problems for competitions in schools or universities, or mathematical puzzles for a wide audience. In addition to this core material, you will also find papers on the connections between mathematics competitions and mathematical research, matters relating to the teaching of mathematics and presentations of specific competitions as they are organized in specific spots all over the world for various well-defined groups of participants. Any interested reader should certainly find something particularly suited to his or her special areas of engagement.

Starting with a brief foreword by Alexander Soifer, whose tenure as President of the WFNMC ended with the Semriach congress, this, the first of the two volumes, is composed of mathematical problems, as they are commonly set at competitions. The interested reader will find a wide selection of fascinating problems, many of which were specially developed for the congress, as well as background on some typical paths that can lead from mathematical research or other mathematical pursuits to the creation of competition-worthy problems. It is our hope that you, the reader, will find these papers and problems as interesting and enjoyable as we, the congress participants did in creating and discussing them. In the second companion volume (available separately), we find material relating to the scientific and organizational background of such problems; the connections between mathematics competitions, research and teaching and some special specific national and international competitions.

If you are interested in more material of this type, you might also be interested in *Mathematics Competitions*, the journal of the WFNMC. The journal is published online twice annually, and old as well as current issues can be found at www.wfnmc.org/journal.html.

Part 1

Some Paths Leading from Interesting Mathematics to the Development of Potential Competition Problems

There are many ways to develop interesting and original problems in elementary mathematics. One of the main interests of all participants in the congresses of the WFNMC, or really the defining commonality of all people working on any mathematics competitions at a national or international level, is the intense interest in finding new problems whose answers aren't immediately obvious, but can be found with a certain reasonable amount of work. Such problems can be side-products of mathematical research, variants of previously known problems, or just witty logical observations that lead to a question.

In this part, we find seven such paths to competition-style problems, as presented at the WFNMC8 congress. The paths presented here are quite disparate, but all lead to amusing, fascinating problems, ranging in levels of difficulty from fairly easy to quite advanced. It is hoped that the interested reader will enjoy tackling the problems as they are presented. Most problems are presented with solutions (where they are known. . .), but a big part of the fun in dealing with this type of "puzzle math" lies in tackling the challenge of finding a solution on your own, always in the knowledge that a complete solution is readily available for perusal whenever the search for a solution turns out momentarily to be a bit frustrating. If you cannot find the solution to a certain problem and none is offered here, you might enjoy using this as an excuse to get in touch with the author of the chapter in question!

Chapter 1.1

Some Standard-Like Problems and Non-standard Solutions

Krzysztof Ciesielski

Jagiellonian University, Kraków, Poland

1. Introduction

If someone is asked for a one-word description of some particular mathematical competition problem, he or she may offer a descriptive term like "geometry" or "inequality" as an answer. On the other hand, a typical reply could also be "easy" or "difficult". However, we may also look at the problem from a completely different point of view. Sometimes, the formulation of a problem can be rather simple, but the solution may nevertheless turn out to be either unexpected or to require certain nonstandard tricks. Generally, such problems are of special interest for mathematicians involved in mathematical competitions, and several problems of this kind are presented in this chapter. (Note that many more such problems are presented in Ciesielski (2019)).

Interesting problems are commonly passed on from one mathematician to another, and, in many cases, it is almost impossible to verify who is the original author of a particular problem. Some of them finally become elements of the so-called mathematical folklore (like the famous problem of the existence of a monochromatic triangle in the complete graph K_6 colored by two colors, see, for example, Soifer (2009)). They may be published in several problem books, or appear in web pages. However, in books, the authors usually present the problems (and, above all, the solutions) in their own individual manner. This will be the case in this chapter as well.

The chapter is organized as follows. In Section 2, the problems are stated, and their respective solutions are then presented in Section 3. When a problem and its solution are placed close to each other in print, a reader may unwillingly have a look at the solution before having had the opportunity to give it any thought, and, for many mathematicians, such a glimpse can be enough to solve the problem. For the convenience of the reader, the problems and their solutions are therefore presented seperately.

Most of the problems presented here are absolutely elementary.

2. Problems

We start with two problems concerning probability.

Problem 2.1. A lottery is advertised with the slogan "every third ticket wins". A ticket costs 2 euro. In fact, however, every third ticket results in neither a win nor a loss; out of 100 tickets available, 33 yield a "win" of 2 euros. If I have such a ticket, I do not actually receive 2 euros, but rather am allowed to draw another ticket. There are therefore 100 tickets available at the sales kiosk, of which one is actually a winning ticket, and 33 of which are "winning 2 euros" tickets. I invest 2 euro in a ticket. What is the probability that I will win?

Most frequently, anyone with basic knowledge about probability answers that a solution requires terrible calculation and suggests a classical calculation concerning conditional probability. Indeed, the probability $P(A_1)$ that I win in the first attempt is equal to $P(A_1) = \frac{1}{100}$, the probability that I win in the second attempt is equal to $P(A_2) = \frac{33}{100} \cdot \frac{1}{99}$, the probability that I win in the third attempt is equal to $P(A_3) = \frac{33}{100} \cdot \frac{32}{99} \cdot \frac{1}{98}$, and so on. To get the required probability, we have to add $P(A_1) + P(A_2) + \cdots + P(A_{34})$. Even some mathematicians specializing in probability theory have answered this way.

This problem has a very simple solution, different from the one presented above, and the challenge is to find this simpler solution.

The next problem also appears at first glance to be a problem requiring enormous calculations in the solution.

Problem 2.2. We put n letters into n addressed envelopes: one letter to one envelope. Let p_k be the probability that precisely k letters were put into their proper envelopes. Prove that for $n \geq 100$, the following inequality holds:

$$p_0 \cdot p_1 \cdot p_2 \cdot \ldots \cdot p_n \leq \log \frac{\sqrt{2\pi n}}{2\pi n!} e^{-\pi n}.$$

Hint. Recall Stirling's formula: for each natural number n there is a $\theta \in (0, 1]$ such that $n! = n^n \times e^{-n} \sqrt{2\pi n} e^{\frac{\theta}{12n}}$.

We now turn to some problems concerning numbers.

Problem 2.3. Prove that $\frac{1}{2} \cdot \frac{3}{4} \cdot \frac{5}{6} \cdot \ldots \cdot \frac{99}{100} < \frac{1}{10}$.

Problem 2.4. A box was filled with 2018 empty boxes. In turn, the next 2018 boxes were put into some of these 2018 boxes (i.e. 2018 new empty boxes were placed into each of the chosen boxes). The procedure was continued for some time. How many boxes remained, if the number of boxes containing other boxes is now 2018?

Problem 2.5. Do 2018 positive consecutive integers exist, among which there are exactly seven primes?

The next problems are connected with "real-life problems", although the situation described in the problem is sometimes rather unreal.

Problem 2.6. On a circle, n long poles are placed at equal distances from each other. A mathematical dwarf sits on each of these poles. Every 5 minutes, two dwarfs change the pole on which they are sitting by jumping to the pole next to it (we do not know which of the two neighboring poles they jump to). For what values of $n \leq 2018$, could it happen that, after some time, all the dwarfs will be sitting on the same pole?

Here, we assume that mathematical dwarfs are so small that they do not have any problem with gathering together on one pole. Moreover, they do not drink and do not eat. They only jump and think about mathematics. They may live infinitely long.

Problem 2.7. Four tourists met at night at the bank of the river in front of an old narrow bridge. The bridge has some holes in it, so they need a torch to see where they are going. Moreover, not more than two persons may be on the bridge simultaneously. The tourists unfortunately only have one torch, so they have to get to the other side by the following method: two persons cross with the torch, one comes back with the torch, two persons cross with the torch, and so on. The tourists cannot cross the bridge at equal speeds. Tourist A needs 1 minute to cross the bridge, tourist B needs 2 minutes, tourist C needs 5 minutes, and tourist D needs 10 minutes. Of course, if two of them cross with one torch, they must walk at the speed of the slower one. Find the smallest time in which all four can find themselves on the other side of the river.

A typical answer obtained to this question is 19 minutes. The method to obtain 19 minutes is as follows: A and B go to the other side, A comes back; A and C cross, A comes back; A and D cross. In this way, we get $2 + 1 + 5 + 1 + 10 = 19$. However, one may guess, that if this puzzle is being mentioned here, the correct solution must be different. Indeed, this is the case.

Problem 2.8. Every day, at exactly 12 noon, a ship starts its journey from Naples to Amsterdam. It travels for exactly 7 days, and then arrives in Amsterdam. There, it meets a ship starting in the opposite direction, because each day, exactly at 12 noon, a ship also starts its journey from Amsterdam to Naples. Each ship sails precisely 7 days. This has been happening for many years. Ships always take the same route. If I board a ship in Amsterdam, how many ships sailing in the opposite direction will I meet during the trip? Meeting in the port, at the very beginning and at the end of the journey counts as well.

Many people answer quickly: eight. Reportedly, many mathematicians at one conference gave this answer. Unfortunately, it is wrong.

Problem 2.9. We have five volumes of the Great Mathematical Encyclopedia standing on a bookshelf. Unfortunately, on the first page of the first volume there is a mathematical beetle which loves to gnaw through books. It takes 2 hours for this beetle to gnaw one cardboard cover, and one hour for it to gnaw through all the interior pages of one volume. The beetle immediately went to battle in the direction of volume five. He started from the first page of the first volume. How long will it take him to reach the last page of volume five?

This calculation seems to be simple. Three volumes will be eaten in 5 hours each ($2 \cdot 2 + 1 = 5$), for the first volume we have $3 = 1 + 2$, as one cardboard cover is not gnawed, and we get the same number $3 = 1 + 2$ for volume five, so the answer is 21. However, once again, this is not a correct answer. Here the puzzle is really tricky.

Now, we turn to some geometrical problems.

Problem 2.10. The angles accompanying vertices P and Q in a quadrilateral are right, and the lengths a and b are given (see Figure 2.1). The sides marked c are equal, but the length c is unknown. Determine the length of line segment PQ.

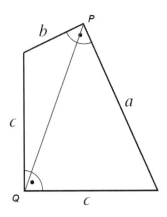

Figure 2.1.

Problem 2.11. The ratio of the internal angles of a triangle is 1:5:6. The length of the longest side is 6. What is the length of the altitude on this side?

We finish with a story concerning a problem which became famous in some mathematical communities a long time ago.

Problem 2.12. A rectangle is the union of pairwise disjoint rectangles, each of which has at least one integer side. Prove that the big rectangle has at least one integer side.

Unlike the other problems in this chapter, we will present the solution to this problem right away.

Consider a rectangle R placed in the coordinate plane, and let the sides of R be parallel to the axes, so $R = [a, b] \times [c, d]$. Now, find a double integral of a function f given by $f(x, y) = e^{2\pi i(x+y)}$ over R. Using the rules of integration and the Fubini theorem, we have

$$\iint_R e^{2\pi i(x+y)} dx dy = \int_a^b \left(\int_c^d e^{2\pi ix} \cdot e^{2\pi iy} dy \right) dx$$

$$= \int_a^b e^{2\pi ix} \left[\frac{1}{2\pi i} \cdot e^{2\pi iy} \Big|_c^d \right] dx$$

$$= \frac{1}{2\pi i}(e^{2\pi id} - e^{2\pi ic}) \int_a^b e^{2\pi ix} dx$$

$$= \frac{1}{2\pi i}(e^{2\pi id} - e^{2\pi ic}) \left[\frac{1}{2\pi i} \cdot e^{2\pi ix} \Big|_a^b \right]$$

$$= \frac{1}{-4\pi^2}(e^{2\pi id} - e^{2\pi ic}) \cdot (e^{2\pi ib} - e^{2\pi ia}).$$

So,

$$\iint_R e^{2\pi i(x+y)} dx dy = 0$$

if and only if

$$(e^{2\pi id} - e^{2\pi ic}) \cdot (e^{2\pi ib} - e^{2\pi ia}) = 0,$$

which holds if and only if $d - c$ or $b - a$ is an integer, as the complex function given by $g(t) = e^{it}$ is periodic with the fundamental period 2π. Thus, $\iint_R e^{2\pi i(x+y)} dx dy = 0$ if and only if the rectangle has at least one integer side.

Now, let T denote the big rectangle and let $T_1, T_2, \ldots, T_n$ denote the small rectangles. As each of the T_i has at least one integer side, we have $\int_{T_i} f(x, y) dx dy = 0$ for each i. But

$$\int_T f(x, y) dx dy = \sum_{i=1}^n \int_{T_i} f(x, y) dx dy$$

and $\int_T f(x, y) dx dy = 0$, which means that the big rectangle has at least one integer side.

This proof is short and beautiful, but very advanced. It uses the double integration of complex functions and advanced results like the Fubini theorem, although the formulation of the problem is absolutely elementary. This problem was presented at several conferences devoted to advanced mathematics, where it raised a great deal of interest among the participants, but at the time of those conferences, no one was able to supply an elementary proof. Does an elementary solution exist?

3. Solutions

In this section, we present the solutions of the problems given in Section 2. The solution of each problem with the number 2.n has the number 3.n.

Solution 3.1. The answer is $\frac{1}{67}$ and may be obtained very quickly. It is sufficient to note that tickets which "draw a tie" need be not taken into account at all, independent of how many of 33 tickets were involved. We therefore have 67 tickets, one of which wins, and 66 of which lose. This solution is really much quicker than a "classical" one concerning conditional probability.

Solution 3.2. In this problem, there is a big "trap". Quite frequently in mathematics, excess information not only does not help, but even causes trouble. Some problems concerning real numbers are much easier to solve when they are stated in a general form for abstract spaces, giving only the assumptions actually needed in the problem. This is the case in Problem 2.2. The Stirling formula is not required at all. Also, the inequality is true for all $n \geq 2$. It is enough to note that $p_{n-1} = 0$. Indeed, if $n-1$ letters are in the proper envelopes, then the nth letter will be in the proper envelope as well. Thus, for $n > 1$, one of factors of the product $p_0 \cdot p_1 \cdot p_2 \cdot \ldots \cdot p_n$ is equal to 0, so the product is equal to 0 as well. Most people calculate the individual probabilities p_k for $k = 0, 1, 2 \ldots$ when they are asked this question, which is definitely not a pleasant task.

Solution 3.3. Even if the use of computers or calculators is not allowed, one option is to actually perform the calculation. If one decides to do this, it turns out that we get required inequality quite quickly, as $\frac{1}{2} \cdot \frac{3}{4} \cdot \frac{5}{6} \cdot \ldots \cdot \frac{63}{64}$ is already smaller than $\frac{1}{10}$. Nevertheless, it would be nice to find a more interesting mathematical solution.

The given inequality is equivalent to

$$\left(\frac{1}{2} \cdot \frac{3}{4} \cdot \frac{5}{6} \cdot \ldots \cdot \frac{99}{100} \right)^2 < \left(\frac{1}{10} \right)^2,$$

which is equivalent to

$$\frac{1 \cdot 1 \cdot 3 \cdot 3 \cdot 5 \cdot 5 \cdot \ldots \cdot 99 \cdot 99}{2 \cdot 2 \cdot 4 \cdot 4 \cdot 6 \cdot 6 \cdot \ldots \cdot 100 \cdot 100} < \frac{1}{100}.$$

We may write this last inequality in another form, namely

$$\frac{1\cdot 3}{2\cdot 2}\cdot\frac{3\cdot 5}{4\cdot 4}\cdot\frac{5\cdot 7}{6\cdot 6}\cdot\ldots\cdot\frac{97\cdot 99}{98\cdot 98}\cdot\frac{99}{100}\cdot\frac{1}{100} < \frac{1}{100}.$$

Now, note that each of the factors (not taking into account the last one) is smaller than 1, as each (except $\frac{99}{100}$) is of the form $\frac{(n-1)\cdot(n+1)}{n\cdot n} = \frac{n^2-1}{n^2}$, and so the whole product is smaller than $\frac{1}{100}$.

Solution 3.4. Here, the solution is based on a clever observation. If the number of boxes containing other boxes is 2018, then filling the boxes was done 2018 times, so 2018 boxes were filled and $2018\cdot 2018 + 1$ boxes were used. Thus, the number of empty boxes is equal to $2018\cdot 2018 + 1 - 2018 = 2018^2 - 2017$.

Solution 3.5. We first consider another problem:

Do 2018 positive consecutive integers exist, among which there are no primes?

This problem is easier than the previous one. It is not surprising that such numbers exist. It is enough to consider the sequence $2019! + 2$, $2019! + 3, \ldots, 2019! + 2019$. These consecutive numbers are divisible by $2, 3, \ldots, 2019$, respectively.

Now, we turn to the solution of the genuine problem. Consider the first 2018 natural numbers: $1, 2, 3, 4, \ldots, 2018$. Of course, there are far more than 7 prime numbers among them, as this sequence includes the numbers $2, 3, 5, 7, 11, 13, 17, 19, \ldots$ Now, take away the first number, i.e. 1, and write the next consecutive number, i.e. 2019 at the end of the sequence. Then the number of primes in our sequence does not change, as 1 is not prime and 2019 is divisible by 3. Next, remove the 2 and add 2020 at the end. Then the number of primes in the resulting sequence is reduced by 1, as we threw away one prime and added a composite number. We can then continue in this manner. After each step, the number of primes either will not change or will be enlarged by 1 or will be reduced by 1. As there were more than seven primes in the sequence at the beginning, and there are none left at the end, there must have been precisely 7 primes in the sequence at some intermediate point.

This reasoning may be regarded as a kind of discrete version of the Intermediate Value Theorem (Darboux Property). The method used here is known as the method of invariants. This method is used also in the solution of the next problem.

Solution 3.6. Number the poles $1, 2, \ldots, n$ as they are placed around the circle, such that pole 1 has pole 2 and pole n as its neighbors.

First, assume that n is odd. In this case, the solution is simple. In the first step, dwarfs sitting on the poles numbered 2 and n jump to pole 1. In the next step, a dwarf occupying pole 3 jumps to pole 2, a dwarf from pole $n - 1$ jumps to pole n. This procedure continues and after some time, all dwarfs will be on pole 1.

When n is even, the reasoning is slightly more complicated.

First, assume that n is divisible by 4. Split the poles into $\frac{n}{4}$ groups of four. The first group consists of poles 1, 2, 3, 4, the next of poles 5, 6, 7, 8, and so on. Consider the first group. In the first step, a dwarf from pole 4 jumps to pole 3, and a dwarf from pole 2 jumps to pole 1. In the next step, two dwarfs sitting on pole 3 jump to pole 2, in the following step, they jump to pole 1. Thus, these four dwarfs arrive on pole 1. Then, the dwarfs on poles 5, 6, 7, 8 repeat this procedure and find themselves on pole 5. Now, the next four dwarfs go to one pole, and so on. When all the dwarfs are grouped in groups of four on single poles, the dwarfs form pairs and those pairs jump to the neighboring poles in order to eventually reach pole 1.

When n is even, but not divisible by 4, the dwarfs cannot be on one pole at any moment. In order to see this, let us color poles in red and blue alternately, i.e. red, blue, red, blue, ... Then a dwarf jumping to a neighboring pole always changes the color of the pole he is sitting on. In the beginning, the number of dwarfs on a red pole is equal to $\frac{n}{2}$, and so it is odd. After any jump, the number of dwarfs on red poles is either reduced by two (if the jumps are made by two dwarfs sitting on red poles) or is enlarged by two (if the jumps are made by two dwarfs sitting on blue poles) or does not change (if one dwarf jumped from a red pole to a blue pole and the second dwarf jumped from a blue pole to a red pole). Thus, the number of dwarfs sitting on red poles is still odd, and it will remain odd after any number of jumps. In particular, at any moment, at least one red pole is certainly occupied by a dwarf. The same reasoning applies to blue poles, and so there must always be at least two occupied poles at any given time.

Solution 3.7. Generally, most people trying to solve this problem start off with the idea that it should be tourist A who carries the torch alone when going back from the opposite side of the river, as he is the quickest and it should be the least waste of time for him to go back and forth. However, the idea of the correct solution is different. Tourists C and D are the slowest, and their crossing uses up the most time, so they should cross the bridge together. The optimal method of crossing is as follows: A and B cross, A comes back; now C and D cross, B comes back with the torch, and finally A and B cross again. The total crossing of the group then takes $2 + 1 + 10 + 2 + 2 = 17$ minutes. Of course, formally, we also require proof of the fact that it is impossible to cross the river in less than 17 minutes, and an interested reader may like to try his or her hand at constructing such a proof.

Solution 3.8. The answer is 15. People answering "8" do not notice, that when a ship starts in Amsterdam, it meets not only the ships which left Naples on this day and later, but also all the ships which left Naples last week. The first ship from Naples, met in the port, started from Naples one week earlier. If a ship from Amsterdam starts, say, on July 18, then it meets ships that leave Naples between July 11 and July 25. If the ships travel with the same speed, periods between meetings are always 12 hours long.

Solution 3.9. Here, the trick is connected with the position of books on a shelf. Note that a beetle, although mathematical, does not read a book; it gnaws it. Five volumes are placed in the shelf as shown in Figure 3.1.

Figure 3.1.

When the beetle is on the first page of volume 1 and goes to volume 5, it is not interested in the contents of volume 1, but goes directly through the front cover of volume 1 to volume 2, as is illustrated in Figure 3.2.

Figure 3.2.

Thus, it reaches the last page of volume 5 in time $3 \cdot 5 + 2 + 2 = 19$ hours.

In the following two geometrical problems, the solutions are based on a technique of "filling out by drawing". In many cases, when some extra lines are drawn in the picture, the solution then appears to be almost obvious, although it seemed to be very difficult before. The following two problems are representative examples of this method.

Solution 3.10. We may put the point P in the vertex of a square and place segments a and b as parts of two sides of the square containing P (see Figure 3.3). Then Q is a point of the intersection of diagonals of the square.

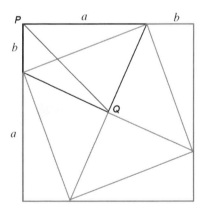

Figure 3.3.

Now, it is easy to note that the length of a side of the square is equal to $a + b$, and segment PQ is half of a diagonal of the square. It follows that its length is therefore equal to $\frac{\sqrt{2}}{2}(a + b)$.

Solution 3.11. If the ratio of the internal angles of a triangle ABC is 1:5:6, then the angles are equal to $15°$, $75°$ and $90°$, and the triangle is therefore right and the side of length 6 is its hypotenuse AC. This hypotenuse is the diameter of the circumcircle of the triangle, with center S. We draw the radii SB and SB' of the circle, where B' is the reflected image of B with respect to the line AC. Furthermore, we also draw the segment BB', as shown in Figure 3.4.

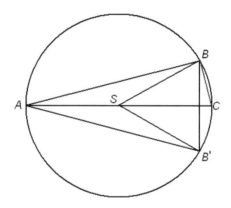

Figure 3.4.

Now, the solution is visible in the picture. The angle BAB' is equal to $30°$ and it is an angle inscribed in a circle. The angle BSB' is a central angle subtended on the same arc as the angle BAB', and is therefore equal to $60°$. The triangle BSB' is an isosceles triangle with equal sides BS and $B'S$, and it is therefore an equilateral triangle and its side has a length equal to the radius of the circle, i.e. 3. It follows that the altitude on the side AC in the triangle ABC is equal to $\frac{3}{2}$.

Solution 3.12. Finally, we come to the rectangle problem. The interested reader will have anticipated that an elementary solution must exist, and we present an idea of how such a solution can be constructed here.

As in the first solution, let T be a big rectangle and $T_1, T_2, \ldots, T_n$ be small rectangles. Consider the plane with Cartesian coordinates. Put the rectangle T in the plane so that its lower left vertex is in the origin. Draw lines parallel to the axes: $\{\frac{n}{2}\} \times R$ and $R \times \{\frac{n}{2}\}$ for $n \in Z$. Now, color the obtained squares to get an "infinite chessboard" (see Figure 3.5). Assume that the square $[0, \frac{1}{2}] \times [0, \frac{1}{2}]$ is black.

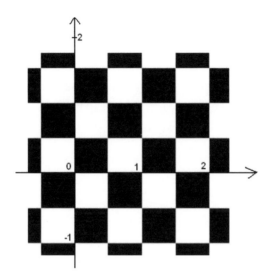

Figure 3.5.

Consider any rectangle with sides parallel to the axes. It is easy to notice that if it has at least one integer side, then the area of its black part is equal to the area of its white part. This means that the "black area" of each of the rectangles $T_1, T_2, \ldots, T_n$ is equal to its "white area". Therefore, the "black area" of T is also equal to its "white area".

Let the lengths of the sides of T be x, y with x placed horizontally. Suppose that neither of the numbers x, y is an integer. Consider a new rectangle T' also placed in the plane with its lower left vertex in the origin with sides $x - \lfloor x \rfloor$, $y - \lfloor y \rfloor$, where we let $\lfloor z \rfloor$ denote the integer part of z (see Figure 3.6). Simply put, T' is made from T by cutting off some rectangles with integer sides. Of course, $x - \lfloor x \rfloor \in (0, 1)$ and $y - \lfloor y \rfloor \in (0, 1)$. If the "black area" of T is equal to its "white area", then this is also the case for T'. It is easy to notice, however, that if the sides of T' are $x - \lfloor x \rfloor \in (0, 1)$ and $y - \lfloor y \rfloor \in (0, 1)$, then the "black area" of T' is greater than the "white area", a contradiction.

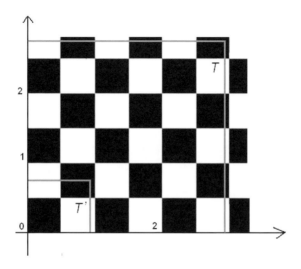

Figure 3.6.

The funny thing in this solution is that it is mathematically the same solution as presented in Section 2. The only difference is in the integrated function. Here, we just integrate the characteristic function of the region colored black, $\chi_B : R \to \{0, 1\}$.

In fact, it turns out that this problem has many different solutions. Several of them are presented in Wagon (1987), see also Shen (1997).

Bibliography

Ciesielski, K. (2019). 102 Math Brainteasers for High School Students, New York, Tom eMusic.

Shen, A. (1997). Mathematical entertainments. *The Mathematical Intelligencer* 19(1), 12–14.

Wagon, S. (1987). Fourteen proofs of a result about tiling a rectangle. *American Mathematical Monthly* 94(7), 601–617.

Soifer, A. (2009). *The Mathematical Coloring Book*. New York: Springer.

Chapter 1.2

Balls and Polyhedra

Robert Geretschläger

BRG Kepler, Graz, Austria

1. Introduction

The study of polyhedra and their various properties is a never ending source of fascination. This is certainly reflected in the many polyhedron-themed problems posed at mathematics competitions all over the world, even if it is true that the study of most aspects of solid geometry is still very much in decline in many school systems. In this chapter, I plan to concentrate on one particular aspect of polyhedra that is perhaps not quite so commonly used in competition problems, namely the "roundness" of polyhedra.

What exactly do I mean by this? One somewhat naïve way of putting it is the question of how well specific polyhedra approximate a sphere, but this is not a trivial property to make it mathematically precise. I will explain a few attempts at making this concept more precise in Section 3. For now, suffice to say that sufficiently "round" polyhedra can be used to produce spherical balls, and balls are certainly a familiar topic for pretty much all participants in mathematics competitions.

I have long been interested in the general topic of polyhedra, but my long-standing (and until recently somewhat hibernating) interest in the topic was rekindled by an observation I made a while back at a local fitness studio. An advertisement for the studio was printed on a ball prominently situated near the entrance, as shown in Figure 1.1.

As is often the case with such equipment, this ball is sewn together from patches of leather, and two of these patches are in the form of regular

Figure 1.1.

heptagons. As it happens, the regular heptagon is also a specific figure I have been interested in in the past (see Geretschläger, 1997) and from this and my previous work on numerical properties of polyhedra (see Geretschläger, 2002) I was immediately aware of some of the more uncommon implications of heptagonal patches on such a ball.

Of course, I was immediately hooked again. Investing some thought into the properties of the heptagonal prism associated with this ball (the ball can be thought of as resulting from uniform inflation of a sufficiently elastic prism of this type, as is indicated in Figure 1.2) had me diving right back into the world of polyhedra, this time with my eyes firmly focused on the types of polyhedra that have something to do with balls typically used for specific sporting activities, as well as their many variants.

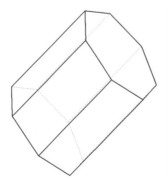

Figure 1.2.

2. Some Possible Competition Problems

Since the main topic at hand is the consideration of mathematical content appropriate for competition problems, I would like to start out by suggesting a few problems from this area that could be used in competitions of various types and at various age levels, along with their solutions. A lot of the mathematical content involved in devising the solutions of these problems will be of further interest at later stages of this chapter.

Here are a few possible competition problems dealing with balls and polyhedra.

Problem 2.1. Joe has an unlimited supply of leather patches in the form of regular pentagons and regular hexagons. All of these have edges of the same length. He wishes to use these patches to sew balls together. The only requirements for his balls are that they must be completely closed (i.e. inflatable) and convex. They can be of any size.

(a) Determine all regular balls that Joe can make using the patches in this way. (Note that a "regular" ball is one in which the same types of polygons meet in each of its vertices.)
(b) Determine all irregular balls that Joe can make using the patches in this way.

Solution to Problem 2.1. We first note that Euler's formula must certainly hold for any convex polyhedron, i.e. we have $v + f = e + 2$, where v denotes the number of vertices, f the number of faces and e the number of edges. If we let p denote the number of pentagons Joe uses to make a certain ball and h the number hexagons he uses for this ball, we obtain $p + h = f$.

Since each pentagonal face has five sides and each hexagonal face has six, and each edge of the polyhedron is shared by two faces, we have $5p + 6h = 2e$, and similarly, since each vertex must be shared by three faces (there can never be more, since each interior angle of a face is at least 108°, and the sum of the interior angles of the faces meeting in a vertex must be less than 360°), we also have $5p + 6h = 3v$. Multiplying Euler's formula by a factor of 6, we obtain

$$6v + 6f = 6e + 12,$$

and substituting for f, $2e$ and $3v$, respectively, gives us

$$2(5p + 6h) + 6(p + h) = 3(5p + 6h) + 12,$$

or $p = 12$. We see that any ball, regular or not, that Joe can make, will certainly have 12 pentagonal faces.

(a) Turning our attention now to the vertices of any possible regular balls that Joe can make, we note that his options are quite limited right from the beginning. We have already noted that there must always be three faces meeting in each vertex. These three cannot all be hexagons, since the internal angles of the hexagons are all equal to 120°, and $3 \times 120° = 360°$. Three regular hexagons with a common corner therefore must all lie in a common plane and cannot form a vertex of a convex polyhedron. This leaves only three options; there can be three pentagons meeting in each vertex of one of Joe's polyhedra, or two pentagons and a hexagon, or two hexagons and a pentagon.

In the first case, Joe obtains a regular dodecahedron, as he is assembling 12 regular pentagons to make a closed convex polyhedron (Figure 2.1).

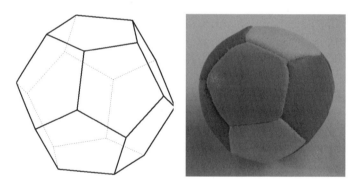

Figure 2.1.

If two hexagons and one pentagon meet in each vertex, Joe obtains a truncated icosahedron; i.e. the classic soccer ball, as shown in Figure 2.2.

Each pentagon must be surrounded by five hexagons in this case, and any two hexagons sharing a common edge must also share a common vertex with a pentagon at either side of that common edge. For reasons of symmetry, this means that the truncated icosahedron is the only possible ball Joe can make in this case.

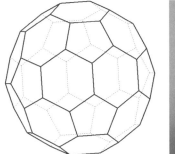

Figure 2.2.

This leaves us with the question of whether Joe can make a ball with two pentagons and a hexagon sharing each of the vertices. This turns out not to be possible. Since no two hexagons could share a common edge in such a polyhedron, each hexagon would have to be surrounded by six pentagons, as illustrated in Figure 2.3.

Figure 2.3.

Each of the two pentagons sharing a common edge in this configuration would then share each of their common vertices with a hexagon. This means that there would have to be two hexagons sharing each of the other pentagonal vertices with a pentagon, yielding a contradiction. Such a ball can therefore not exist.

(b) We now turn our attention to the matter of irregular balls that Joe can make. The object shown in Figure 2.4 seems to suggest that such an object can exist, composed of 12 pentagons and 10 hexagons.

Figure 2.4.

Unfortunately, this is, in fact, not possible. (The polygons in Figure 2.4 are not plane figures despite the fact that they may appear to be so at first glance.) In order to prove this, we must simply consider some of the symmetries that an object of this type would have to embody.

If any of Joe's polyhedra has a vertex in which three pentagons share a common vertex, the entire polyhedron must be the regular dodecahedron.

Since the right-hand pentagon in Figure 2.5 has a common side with each of the other two pentagons, there exists another pentagon symmetric to the one with respect to the bisector of the common side of the other two pentagons, which also has a common side with both of them. In other words, the two left-hand sides of the pentagons enclose an angle of 108°. Only a regular pentagon can be fit into the space in such a way that it will share a side with both pentagons, and this same forced angle will appear in every

Figure 2.5.

vertex, meaning that the completion of the only possible closed polyhedron with this trio of pentagons in a vertex must be a regular dodecahedron, as claimed.

Similarly, if any of Joe's polyhedra has a vertex in which two hexagons and a pentagon meet, the analogous symmetry argument shows us that every vertex must have this property, and we once more end up with the truncated icosahedron.

This leaves us with only the option that there exist no vertices of either type. In this case, two pentagons and a hexagon must meet in every vertex, but we have already argued in part (a) that no such a polyhedron can exist, and we must come to the conclusion that Joe cannot make any irregular polyhedra out of his patches at all.

Having considered this problem closely related to the classic soccer ball at some length, it is interesting to note that we can also ask a question about a quite different type of ball with almost the same geometric content.

Problem 2.2. The surface of any modern golf ball is covered in so-called dimples. These are small circular indentations that optimize the aerodynamic properties of the golf ball in flight. A certain common type of dimple pattern has the property that any individual dimple is always surrounded by either exactly five or exactly six other dimples. Prove that the number of dimples surrounded by exactly five others is independent of the total number of dimples on the golf ball, and determine this number (Figure 2.6).

Figure 2.6.

Solution to Problem 2.2. We already know from the first part of the solution to Problem 1 that the number of such dimples must be 12. The surface of the golf ball can thought of as a polyhedron whose faces are all pentagons or hexagons, with three of these polygons meeting in each vertex, is we simply imagine the planes in which the circles of the dimples lie as the planes of the polygons. Their intersections are then the edges and vertices of the polyhedron, and we have already shown that the number of pentagonal faces must always equal 12, independent of the number of hexagonal surfaces.

Of course, much more elementary questions can be asked about the truncated icosahedron.

Problem 2.3. A classic soccer ball is sewn from leather patches in the shape of regular pentagons and regular hexagons. Each hexagon is sewn along an edge together with three pentagons and three hexagons. Each pentagon is sewn together with a hexagon along each of its edges. There are 12 pentagons on the ball. How many hexagons are there on the ball?

(A) 8   (B) 12   (C) 20   (D) 24   (E) 36

Solution to Problem 2.3. Of course, we already know that the answer is 20, since we have given a great deal of thought to this polyhedron in the previous problems. This is such an elementary problem, however, that it can be solved in a much simpler way. It is not necessary to know anything about Euler's formula; simple double-counting of the number of edges of the soccer ball is enough. This makes the problem accessible to students at a much younger age.

One way to count the number of edges is to note that each pentagon has five sides and each hexagon has 6. Each edge of the ball is a common side of two of these, and letting $p = 12$ denote the number of pentagons and h the (unknown) number of hexagons, the number of edges is therefore equal to $\frac{1}{2}(5p + 6h)$. On the other hand, each of the edges is a side of either one or two of the hexagons. Three of the sides of each hexagon are unique to this one (and shared with a pentagon), while the other three are shared with another hexagon. The number of edges of the ball is therefore also equal to $3h + \frac{3}{2}h$. Since both methods of counting must yield the same result, we obtain

$$\frac{1}{2}(5p + 6h) = 3h + \frac{3}{2}h,$$

which is equivalent to $3h = 5p$, and since we are given $p = 12$, this therefore yields $h = 20$, as stated.

Of course, some interesting problems for an elementary level result by changing just one of the constants. Instead of pentagons and hexagons, the patches can be in the shape of some other kinds of regular polygons. Some examples of problems resulting from this idea are the following.

Problem 2.4. Otto wants to sew his own soccer ball. For this purpose, he cuts a number of equilateral triangles of equal size out of leather and sews them together in such a way that four triangles meet in each corner of the resulting (closed) ball. When he inflates the ball, the leather stretches a bit, and the result is a sphere with radius 1 dm. What is the visible area of one of the triangular patches when the ball is inflated?

(A) $\pi \, \text{dm}^2$ (B) $\frac{3}{4}\pi \, \text{dm}^2$ (C) $\frac{2}{3}\pi \, \text{dm}^2$ (D) $\frac{1}{2}\pi \, \text{dm}^2$ (E) $\frac{1}{3}\pi \, \text{dm}^2$

Solution to Problem 2.4. The total surface area of a sphere with radius 1 dm is, of course, 4π. If four triangular patches meet in each of the vertices, the complete polyhedron can be thought of as a double square pyramid (i.e. a regular octahedron). This means that the ball is made up of eight such triangles, and the area of each is one-eighth of the total surface area of the ball, or $\frac{1}{2}\pi \, \text{dm}^2$.

Problem 2.5. A spherical ball is sewn from leather patches. Five of these patches are square and t are triangular. Which of the following is not a possible value of t?

(A) 2   (B) 3   (C) 4   (D) 6   (E) 12

Solution to Problem 2.5. Each of the square faces of the ball has four sides and each of the triangular faces has three. Since two sides always join to form an edge of the polyhedral ball, the number of edges (certainly an integer) is equal to $\frac{1}{2} \cdot (5 \cdot 4 + 3 \cdot t)$. If t is odd, this is not an integer. It therefore follows that the value of t cannot be 3.

Another idea is to consider the typical tennis ball. This type of sphere is not really polyhedral in the same sense as the other balls we have been looking at, but in a way, it is closely related to an inflated version of the cube. We can then think of the seam of the tennis ball as being an inflated version of the dotted line shown in Figure 2.7(a).

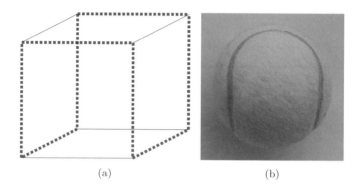

(a)   (b)

Figure 2.7.

Problem 2.6. Xanthippe and Yanni are playing a game with a tennis ball and a pen. Yanni wants to draw as many stars as he can on the ball, but at least two. Xanthippe will then take the pen and draw a continuous curve on the ball connecting two of the stars. If she can draw the curve without crossing the white seam, she wins. If not, Yanni wins. What is the largest number of stars that Yanni can draw and be certain of winning?

(A) 2   (B) 3   (C) 4   (D) 6   (E) 8

Solution to Problem 2.6. The seam divides the tennis ball into two distinct sections. If Yanni draws more than 2 stars, two of them must be in the same sector due to the pigeon-hole principle, and Xanthippe will certainly win. The largest number of stars Yanni can draw if he wants to be sure of winning is therefore 2.

3. Typical Balls and Their Polyhedral Alter Egos

There are many different types of balls, designed for various purposes. Many, like squash balls, are really just spheres from a polyhedral point of view, and these will not be considered here. The reason that many balls have an intrinsic polyhedral structure is that they were originally designed to be sewn together from more or less plane pieces of leather or some other material, and these are the balls we will be focusing on.

Of course, there are many different interesting geometric structures we can find on balls, but there are four extraordinarily common polyhedral structures we can observe on a large majority of balls, and these are illustrated in Figure 3.1. They are represented by the soccer ball, the beach ball, the basketball and the volleyball, respectively.

Figure 3.1.

Let us take a closer look at these four types of balls and the polyhedral structures they represent.

The Soccer Ball

When we are thinking about balls and associated polyhedra, the type of ball that comes to mind first and foremost for most people who think about such things at all is the classic soccer ball. In its geometric essence, this ball is simply a specific Archimedean solid, the truncated icosahedron. Problems 2.1 and 2.3 in Section 2 deal with some of the properties of such balls. Their geometric properties are well documented in the literature on semi-regular polyhedra.

As a soccer ball, this form was really only ever "standard" in the sense that the official World Cup balls were of this form, from 1970 to 2002. Before 1950, World Cup balls had laces, and from 1950 to 1966, the geometry of the World Cup balls had more in common with the type of ball we now associate with volleyball than any Archimedean solid.

What is so special about this specific solid? Taking a closer look at the geometric properties, we find that several desirable properties for a soccer ball are combined in this form:

- It is very regular. As an Archimedean solid, each of the vertices is locally congruent. This means that the ball will roll in a straight line.
- It is very round. There are many possible criteria for roundness, and we will be visiting some good candidates in Section 3. For now, we can note that the smallest minimum of an interior angle in any vertex is 108°; a very large value. This means that the ball will roll smoothly.
- It is composed of just two types of regular polygons. This means that it can be manufactured in a reasonably simple way.

None of these criteria are met in an optimal way, of course. Any of the platonic solids are more regular than any Archimedean solid, and since they are composed of just one type of polygon, they are easier to manufacture by this criterion. This solid is not even the "roundest" by some criteria (see, for instance, http://festival.symmetry.hu/videos-symmetry-festival-2013-delft-the-netherlands/pieter-huybers/). Nevertheless, this form has carved itself a niche in our collective psyche, and problems on soccer balls based on this shape can be posed under the assumption that many of the students taking part in a competition will have some naïve picture of the form in their subconscious to refer to.

Sometimes, the regular dodecahedron is used as a simplified version of a soccer ball. This is with good reason, as is illustrated by the three pictures of balls in Figure 3.2. On the left, we see a ball sewn together from 12 pentagons, and on the right, we see a ball sewn from 12 pentagons and 20 hexagons, i.e. a dodecahedron and a truncated icosahedron. In the middle,

Figure 3.2.

we see a ball in which both structures are visible. Starting with the truncated icosahedron and cutting each of the hexagons in half by a diagonal creates large pentagons around each of the smaller original pentagons, and the dodecahedral structure is made apparent.

The Beach Ball

Perhaps even more ubiquitous than the soccer ball is the plastic beach ball. Most such balls are made up of six identical pieces of plastic, welded together along their edges and held together at their ends by circular patches. Such a ball is shown in the upper left of Figure 3.3. A bit of abstraction shows us that this is, in its geometric essence, more or less a hexagonal prism. As illustrated in Figure 3.3, we see that the rectangular sides of the prism correspond to the side panels of the beach ball, and the hexagonal sides correspond to the circular patches of the ball. The ball can be interpreted as an inflated version of this prism.

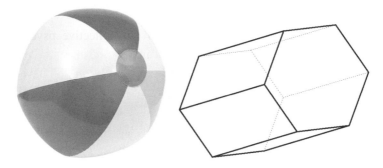

Figure 3.3.

Of course, once we have established that an inflated prism is a good model for a simple plastic ball of this type, there is no reason to restrict ourselves to a hexagonal one. In fact, there are many beach balls on the market, whose structure is based on some other regular prism. One such ball, based on a 12-sided prism, is shown in Figure 3.4.

Figure 3.4.

The Basketball

Unlike the beach ball, the polyhedral structure of the standard basketball is perhaps not quite so obvious. Taking a closer look, we see that its surface is composed of eight panels, each of which is equally large. In fact, as we see in Figure 3.5, four of these panels are identical and the other four are symmetric to these. Each panel has three (curved) sides and three corners,

Figure 3.5.

and four such panels meet in each of six points, three of which are situated on one side of the ball and three on the other.

Somewhat surprisingly, this implies that the standard basketball is essentially an inflated octahedron, made up of eight triangular panels, four of which meet in each of the vertices.

While this octahedron is not completely regular, the fact that the eight "faces" are all either identical or symmetric does mean that the shape is quite regular in some ways. In Figure 3.6, we see how a regular octahedron can be transformed in steps to make the connection a bit more visible. Three of the six vertices of the octahedron wander together on each of the two opposing sides, making the triangular faces long and thin. Introducing curved sides (i.e. inflating the ball and allowing for some elasticity in the panels) then creates the spherical shape.

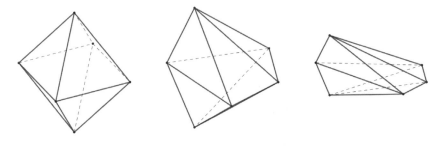

Figure 3.6.

The Volleyball

The structure of the typical volleyball was once even more common than it is now. As already mentioned, this was once also the structure typically used in the manufacture of soccer balls, but was eventually replaced in that context by others. Taking a closer look, we see that the volleyball is made up of two types of strips. Six of these appear to be more or less rectangular, and the other twelve seem to approximate trapezoids. These shapes appear of the surface of the volleyball in groups of three, forming somewhat rounded squares.

We see that the volleyball is essentially an inflated cube in which each of the square faces has been trisected in such a way that the short ends of each of the resulting strips are joined to long sides of strips on the adjacent faces of the cube, as illustrated in Figure 3.7.

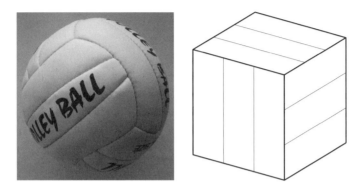

Figure 3.7.

A better polyhedral approximation is obtained by thinking of the outer strips not as trapezoids, but rather as hexagons. This is motivated by the fact that there are in fact six points along each edge of such a panel, in which a corner of another panel is attached. In other words, chamfering the edges of the cube (i.e. truncating the cube's edges) yields a polyhedron composed of 12 hexagons and six rectangles, whose geometric structure illustrates that of the volleyball. Two versions of such a chamfering are shown in Figure 3.8.

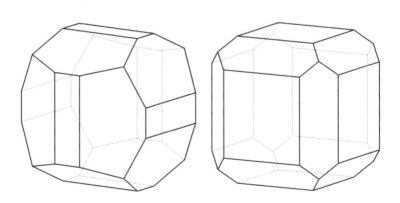

Figure 3.8.

Tennis Balls, Golf Balls and others

Balls with various other types of geometrical surface structure exist, of course. As already mentioned in Problem 2.6, the surface of a tennis ball can be interpreted as being a rounded cube in which four edges have been removed.

Golf balls were mentioned in Problem 2.2, and their dimpled surfaces are typically structured with pentagons and hexagons, although other types are also in use.

Of course, the commonality of all these structures is in the effort to approximate the spherical structure. For some games (squash, table tennis, bowling,...), the balls used are indeed simply spheres, as material is used that can be worked in a continuous fashion. (Note that this is also true for many modern soccer balls.) Certainly, for many games, the essential aspect of a ball is for it to best approximate a sphere, i.e. for it to be as "round" as possible. But what do we mean by this exactly? How can we best measure the "roundness" of an essentially polyhedral structure? Some ideas that can come up in the discussion of this aspect can also lead to interesting geometry and some nice potential competition problems.

4. Criteria for "Roundness"

So, for many sports, we want balls to roll and fly with as little wobble or other irregularity as possible. It is therefore of interest to find criteria for the level of approximation of a given polyhedron to a sphere, since the majority of practical balls are sewn together from plane patches, forming polyhedra before they are inflated to their maximal curvature.

For every polyhedron (in fact, for every closed surface), there exist a circumscribed sphere S_c with minimal volume V_c and an inscribed sphere S_i with maximal volume V_i. For a sphere S, we have $S_c = S_i = S$ and therefore $V_c = V_i$, equal to the volume of the sphere itself. For any polyhedron P, S_i is certainly contained entirely within P, which is itself contained entirely within S_c, and S_i is therefore certainly entirely contained within S_c. This means that $V_i < V_c$ (or, equivalently, $\frac{V_c}{V_i} > 1$) holds for any P. An excellent criterion for the "roundness" of a polyhedron is therefore the value of $r_p = \frac{V_c}{V_i}$. The closer the value of r_p is to 1, the better the spheres S_c and S_i approximate P.

A possible (and reasonably easy) competition problem derived from this concept is the following.

Problem 4.1. We are given a cube C and a regular octahedron O. Which of these has the smallest roundness factor r_p?

Solution to Problem 4.1. Since both of these polyhedra are Platonic solids, they each have both a circumscribed and an inscribed sphere. For the cube with edge-length a, the radius of the inscribed sphere is equal to $\frac{a}{2}$, and the radius of the circumscribed sphere is equal to $\frac{a}{2}\sqrt{3}$, as shown in Figure 4.1.

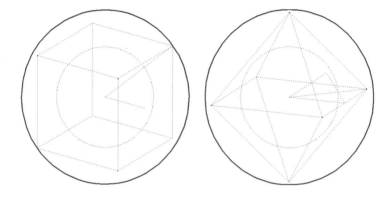

Figure 4.1.

We therefore obtain

$$r_c = \frac{\frac{4}{3}\pi \cdot \left(\frac{a}{2}\sqrt{3}\right)^3}{\frac{4}{3}\pi \cdot \left(\frac{a}{2}\right)^3} = 3\sqrt{3}.$$

For the octahedron with edge-length a, the radius of the circumscribed sphere is equal to $\frac{a}{2}\sqrt{2}$. The radius of the inscribed sphere requires a tiny bit of extra calculation, but denoting the mid-point of the octahedron with O, the mid-point of one of its faces with M and the mid-point of one of the edges of the face containing M with M_a, we see that OMM_a is a right triangle with its right angle in M. The length of OM is the radius r of the inscribed sphere, $OM_a = \frac{a}{2}$ and $MM_a = \frac{1}{3} \cdot h_a$, where h_a has denotes the altitude of the equilateral triangle with edge-length a, i.e. $h_a = \frac{a}{6}\sqrt{3}$.

This gives us $r^2 = \frac{a^2}{4} - \frac{3a^2}{36} = \frac{6a^2}{36}$ or $r = \frac{a}{6}\sqrt{6}$, and therefore

$$r_o = \frac{\frac{4}{3}\pi \cdot \left(\frac{a}{2}\sqrt{2}\right)^3}{\frac{4}{3}\pi \cdot \left(\frac{a}{6}\sqrt{6}\right)^3} = 3\sqrt{3}.$$

We see that $r_c = r_o$, and the cube and the octahedron are equally round by this criterion.

This result is actually a bit surprising. Since an octahedron has eight faces and a cube only has six, we might have expected the octahedron to be "rounder". This is certainly true if we apply the very naïve criterion of number of faces; i.e. calling one convex polyhedron rounder than another if it has more faces. On the other hand, the vertices of the octahedron are certainly "pointier" than those of the cube, and we could think of the cube as being "rounder" by this criterion. We will see how these two solids compare with respect to other possible criteria in a moment. First, however, let us consider one more problem.

Problem 4.2. We are given a cube C and a regular octahedron O. Which of these has a volume that more closely approximates the volume of its circumsphere?

Solution to Problem 4.2. The volume of a cube with edge-length a is, of course, a^3, and the ratio we are looking at is therefore equal to

$$\frac{\frac{4}{3}\pi \cdot \left(\frac{a}{2}\sqrt{3}\right)^3}{a^3} = \frac{\pi}{2} \cdot \sqrt{3}$$

for the cube. The volume of the octahedron with edge-length a is equal to $\frac{\sqrt{2}}{3} \cdot a^3$, and the corresponding ratio for the octahedron is therefore equal to

$$\frac{\frac{4}{3}\pi \cdot \left(\frac{a}{2}\sqrt{2}\right)^3}{\frac{\sqrt{2}}{3} \cdot a^3} = \pi.$$

We see that the octahedron fills its circumsphere to a lesser degree than the cube does, and in this sense, the octahedron is less "round" than the cube.

So far, we have three candidates for roundness criteria for a convex polyhedron P, namely $r_P = \frac{V_c}{V_i}$, the number of faces of P and the ratio of V_c to the volume of P. Considering the criteria involving volume appears to be quite useful, but we must expect the calculation of the volumes involved to be quite cumbersome in general. Counting the number of faces of a polyhedron is easier, but does not really seem to be informative enough. So, what else could we feasibly use?

Let us consider two specific types of polyhedra that we can agree seem to be very "round", and take a closer look at their numerical properties. Perhaps, we will be able to spot some other interesting numbers to work with.

One candidate for such a polyhedron is certainly the truncated icosahedron, i.e. the classic soccer ball. Another is the triangulation of a sphere obtained by marking many points on a sphere in as evenly spaced a manner as possible and then "connecting the dots" to form many triangles. An example of such a polyhedron is shown in Figure 4.2.

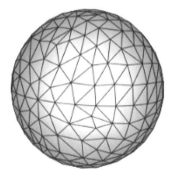

Figure 4.2.

This is reminiscent of the golf ball in Problem 2.2, of course. We can think of a golf ball as representing such a polyhedron if we connect the mid-points of the circular dimples to form the faces of the triangulation. In both cases, we can certainly agree about roundness; there is no doubt that soccer balls and golf balls fit the bill.

A common property of both is the fact that the vertices of neither are in any way "pointy". Another way to say this is that the sum of the interior angles of the faces meeting in each vertex is quite close to 360°. In a truncated icosahedron, each of these angle sums is equal to $120° + 120° + 108° = 348°$, since two regular hexagons and a regular pentagon meet there. In a sphere triangulation/golf ball, we can observe that the triangles meet in each vertex in such a way that the angle sum is maximal. If all triangles were equilateral and six met in each vertex, the angle sum in each vertex would be $6 \cdot 60° = 360°$, which is not possible for a vertex of a convex polyhedron. Just taking a cursory look at the surface of a golf ball does, however, make it seem like six triangles have a common corner in each dimple. In fact, a closer look reveals that some of these vertices will only be common to five triangles, reducing the average face angle sum to less than 360°, as we know from the result of Problem 2.2.

So, here are some candidates for roundness criteria based on face angles:

  (i) R_1: the minimal value of the average angle in each vertex.

  (ii) R_2: the average angle sum per vertex.

  (iii) R_3: the minimum value of the angle sum per vertex.

  (iv) R_4: the average interior face angle over the entire polyhedron.

For the truncated icosahedron, the values of these markers are easily calculated as

$$R_1 = 116°; \quad R_2 = 348°; \quad R_3 = 348° \text{ and } R_4 = 116°,$$

and for a sphere triangulation with a large number of triangular faces, a bit of reflection yields

$$R_1 = 60° - \varepsilon; \; R_2 = 360° - \varepsilon; \; R_3 = 360° - \varepsilon \text{ and } R_4 = 60°,$$

with ε being some small positive value in each case.

We immediately see that R_1 and R_4 are not good candidates for roundness criteria at all, since they do not seem to indicate large values for the sphere triangulation. On the other hand, R_2 and R_3 both seem reasonable at this stage. It seems that a value of either of these markers as close as possible to 360° indicates a type of "roundness" we are seeking to identify. Since the "average" metric of R_2 and the "minimum" metric of R_3 promise to be quite similar, we will only consider one of them in the section to follow, choosing R_2 for now, for reasons of greater ease in the concrete calculations. (An analogous study of R_3 is certain to yield some interesting problems as well, of course.)

Consideration of R_4 does, however, suggest another interesting potential competition problem.

Problem 4.3. We are given a convex polyhedron P. The average size of all interior face angles of P is equal to φ. Prove $60° \leq \varphi < 120°$.

Solution to Problem 4.3. Each face of a convex polyhedron is a convex n-gon. The average interior angle of an n-gon is equal to $\frac{(n-2) \cdot 180°}{n}$, and since $n \geq 3$ is certainly true in each face, we obtain

$$3 \leq n \Leftrightarrow 360° \leq 120° \cdot n = 180° \cdot n - 60° \cdot n$$
$$\Leftrightarrow 60° \cdot n \leq 180° \cdot n - 360° = (n-2) \cdot 180°$$
$$\Leftrightarrow 60° \leq \frac{(n-2) \cdot 180°}{n}.$$

We see that the average interior angle in each face is not less than $60°$, and this is therefore also true for the average interior angle in every face. Equality holds for any polyhedron whose faces are all triangles, such as a tetrahedron.

We now note that the sum of the angles in each vertex is certainly less than $360°$ and the number of faces meeting in each vertex is not less than 3. From this, it follows that the average of the face angles meeting in any vertex is less than $\frac{1}{3} \cdot 360° = 120°$, and this is therefore also true for the average of all interior angles in every vertex, completing the proof.

Another interesting candidate for a roundness criterion results from comparing the number of edges of a polyhedron with the number of its faces or its vertices. (Since these two ideas are mutually dual in nature, we will limit our discussion here to the first option.) This results from some interesting inequalities (see Geretschläger, 2002, Problems 18 and 19) that can also be used as an interesting problem, as follows.

Problem 4.4. Let P be a convex polyhedron with e edges and f faces. Prove

$$\frac{3}{2} \leq \frac{e}{f} < 3.$$

Solution to Problem 4.4. First of all, we would like to show the validity of the left inequality. In order to do this, we note that each face of the polyhedron has at least 3 sides, for a total of at least $3f$ sides. Since each edge is shared by two such sides, we obtain $2e \geq 3f$, confirming the validity of the left inequality.

In order to show the right inequality, we first note that this argument also holds for the e edges and the v vertices of the polyhedron, since at least 3 edges must end in each of the vertices. We therefore obtain the inequality $2e \geq 3v$ in a completely analogous way. Noting that Euler's formula gives us $v = e - f + 2$, we can substitute for v and obtain

$$2e \geq 3(e - f + 2), \text{ or } 3f \geq e + 6.$$

This therefore gives us $\frac{e}{f} \leq 3 - \frac{6}{f} < 3$, completing the proof.

In order to see why this ratio could be a candidate for a roundness criterion, let us take a look at the relevant values for some well-known types of polyhedra.

It seems to be reasonable to assume that a polyhedron with a large number of edges per face is "rounder" than one with less edges per face. This is made plausible by considering the following table:

Object	e	f	e/f
Tetrahedron	6	4	1.5
Cube	12	6	2
Dodecahedron	30	12	2.5
Truncated icosahedron	90	32	2.8125
n-sided prism	$3n$	$n + 2$	$\frac{3n}{n+2}$
n-sided pyramid	$2n$	$n + 1$	$\frac{2n}{n+1}$
Sphere triangulation with k triangles	1.5k	k	1.5

We see that the number of triangles in any sphere triangulation is irrelevant in this context. (This includes the octahedron, icosahedron, etc.) Also, if n is very large, the value of $\frac{e}{f}$ for the n-sided prism approaches 3 as the prism itself gets nearer and nearer to a cylinder. A cylinder is certainly a round object, and we see this is a reasonable concept of roundness of some sort, but not the kind we want. It is also interesting to compare this to the cone, which we would normally consider to a similar type of "roundness" as the cylinder. The cone results from the n-sided pyramid by letting n approach infinity, and we note that the value of $\frac{e}{f}$ for the cone approaches 2. This is interesting to note, since the big "roundness" difference between a cylinder and a cone is the existence of the apex of the cone. This is, in a way, the ultimate non-round point on the surface of a solid, and the difference between a cone and a cylinder, or a pyramid and a prism, is quite marked with respect to this particular number for this reason.

5.   Non-Polyhedral and Non-Spherical Balls

So far, we have been looking only at spherical balls with surface properties relating them to convex polyhedra. While such balls are indeed the most common, it is interesting at this point to take a look at some other types of balls that can also be sewn together from patches.

Old style soccer balls were often made using seemingly strange shapes, like the T-shaped pieces of leather used in the ball shown in Figure 5.1. Three of the T-shaped panels meet in each of the "vertices" of this ball, and there are two types of such "vertices". This actually results in a very interesting and unique type of symmetry.

Figure 5.1.

The structures of the balls in Figure 5.2, on the other hand, are reminiscent of the structure of the volleyball. In the first two balls, there are two parallel strips in each cubic section, and in the third, the three rectangular strips are replaced by more complicated polygonal structures.

As was the case in Figure 5.1, some of the polygons on the balls here are not convex. This results in quite unusual structural properties.

Another interesting type of non-polyhedral ball is made in the same way as the beach ball, but without capping the ends with circular (or polygonal)

Figure 5.2.

pieces. Examples of such balls are the juggling balls shown in Figure 5.3 and the non-spherical balls used for (American) football or rugby, shown in Figure 5.4.

Figure 5.3.

Figure 5.4.

Each of these is composed of four or six strips meeting in common endpoints on either side of the ball. We cannot reasonably interpret such balls as polyhedra, since they would only have two vertices.

Finally, in Figure 5.5, we see some balls with less symmetry than we would expect to be practical for sporting purposes. In fact, these are not typically used for games, but rather for training purposes. The ball on the left is used for rugby training, and is essentially half a standard rugby ball. From a polyhedral point of view, it can be considered an inflated quadratic pyramid, as it is sewn from a (rounded) quadratic panel and four triangular ones. The ball on the right is composed of 66 pentagonal and hexagonal panels, and is used in soccer training to help players gain better control of irregular ball movement.

Figure 5.5.

6. Some Ideas Toward Developing Further Competition Problems

Considering the various properties of balls and polyhedra in the preceding sections has suggested a number of possible ways to derive problems of varying degrees of difficulty that could conceivably be posed at competitions. Many of these simply ask about the existence of polyhedra with specific numerical properties.

Polyhedra are actually close cousins to tessellations in the plane, and in order to find out whether a specific polyhedron exists, it is often useful to first consider a graph with the appropriate properties, and worry about the three-dimensional version in a next step.

For this purpose, Steinitz's theorem is quite useful. This famous theorem states that a graph G is the graph of a convex three-dimensional polyhedron if and only if G is planar and 3-connected. (A graph is called "3-connected" if any pair of vertices remains connected by a path after removal of any two vertices of the graph.)

In Sections 1 and 3, we considered various ideas resulting from Euler's polyhedron formula $v + f = e + 2$. An especially interesting aspect resulted from limiting the types of faces that a polyhedron could have. Specifically, we noted that any convex polyhedron, whose faces are all pentagons and hexagons must have 12 pentagonal faces. An interesting variation results from a slight generalization of this idea.

Let us assume that a convex polyhedron has n faces with a sides each and m faces with b sides each, but no others. This means that we certainly have

$$f = n + m \text{ and } 2e = an + bm,$$

since every edge of the polyhedron is shared by two faces, and assuming that three faces meet in every vertex, we also have $3v = an + bm$. Once again multiplying Euler's formula by 6 and substituting these expressions, we obtain

$$6v + 6f = 6e + 12 \text{ or } 2(an + bm) + 6(n + m) = 3(an + bm) + 12,$$

which simplifies to $6n + 6m = an + bm + 12$.

It is now possible to try out certain specific values for one or the other of these free variables, and check out whether any polyhedra exist for which these numbers apply, and if they do, we can ask more specific questions about them.

A few ideas of this type are suggested by the following:

- Setting $b = 6$ (as before), we obtain the equation $(6 - a) \times n = 12$. It follows that n must be a divisor of 12 greater than 2. It is easy to check that there are no possible polyhedra for $n = 3$. The only possible combinations for n and a are $n = 4$ and $a = 3$, $n = 6$ and $a = 4$ and $n = 12$ and $a = 5$. The latter case was explored at length in Problem 2.1. Some polyhedra resulting from the other cases are shown in Figures 6.1–6.3 . In Figure 6.1, we see a polyhedron with $a = 3, b = 6, n = 4$ and $m = 4$.

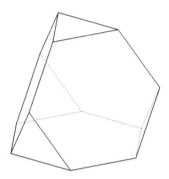

Figure 6.1.

For which other values of m (if any) does there exist a polyhedron with

$$a = 3, b = 6 \text{ and } n = 4?$$

In Figure 6.2, we see polyhedra with $a = 4, b = 6$ and $n = 6$. Their values for m are 2, 3 and 8, respectively.

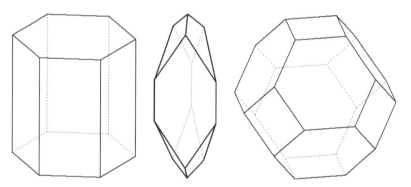

Figure 6.2.

Is it possible to have a polyhedron with 1 hexagonal face and 6 four-sided faces (i.e. $m = 1$)? What about other values of m?

What other special polyhedra exist with only two types of faces?

- Setting $b = 7$, we obtain the equation $(6 - a) \times n = m + 12$. If we could have $m = 1$, m must equal 13, and therefore $a = 5$. Is there such a polyhedron? If so, what does it look like? If not, why not?
- What other interesting values of b can we consider?

Of course, this train of thought can be continued in many ways. An interesting idea would also be to consider non-convex "balls". For instance, we can ask the following question:

- What types of toroidal balls (polyhedral with one hole) exist? What properties can we expect from a reasonably "round" toroidal ball that can be sewn from a small number of patches? (Note that a torus can always be triangulated by many small triangles, yielding a good polyhedral approximation, but such a polyhedron would be very difficult to "sew together".) A picture of a simple toroidal polyhedron, composed of 16 quadrilaterals, is shown in Figure 6.3.

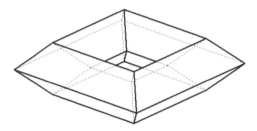

Figure 6.3.

7. Back to the Fitness Studio

Before we end our discussion on polyhedral balls, let us return to the ball we started with, namely the fitness studio ball sewn together from two regular heptagons and seven four-sided patches. Here are a few questions inspired directly by the fitness center ball we could ask our students to solve.

Problem 7.1. Does there exist a convex polyhedron with 2 heptagonal faces and k quadrilaterals as faces with $k \neq 7$? For which values of k does such a polyhedron exist?

Problem 7.2. Does there exist a convex polyhedron with 2 heptagonal faces and $k = 7$ quadrilaterals as faces in which the two heptagons share a side? Does one exist for $k \neq 7$? For which values of k does such a polyhedron exist?

Problem 7.3. Does there exist a convex polyhedron with 2 faces in the form of an m-gon (with $m \neq 7$) and 7 quadrilaterals as sides? For which values of m does such a polyhedron exist?

8. Conclusion

It is a shame that solid geometry has lost so much of its status in most school systems. We do, of course, live in a 3D world, and people's fascination with all things three dimensional can be seen in the current popularity of 3D films and animation. It would seem obvious that people would want to learn more about the intrinsic geometry of the 3D world and three-dimensional shapes. Perhaps, playing with solids and their elementary properties in the way suggested by the problems in this chapter can help a bit toward repopularizing the subject of solid geometry in the future.

Even if these high hopes should come to naught, I would hope that the problems suggested here will prove as entertaining to others as they are to me. Perhaps, we can hope that, at the very least, the world of mathematics competitions can continue to be a fertile ground for this type of problem.

Bibliography

Geretschläger, R. (1997). Folding the regular heptagon, *Crux Mathematicorum with Mathematical Mayhem* 23(2), 81–88.

Geretschläger, R. (2002). Numerical polyhedron problems, *Mathematics Competitions* 15(2), 15–57.

Chapter 1.3

Hunting of Lions: Inversion May Help

Kiril Bankov

University of Sofia and Bulgarian Academy of Sciences, Sofia, Bulgaria

1. Introduction

A mathematical method of hunting a lion: We place a circular cage at a given point of the desert, enter it and lock it. We perform an inversion with respect to the cage. The lion is then in the interior of the cage, and we are outside (Petard, 1938).

This method, while difficult to implement practically, is based on the amazing properties of the geometrical transformation called inversion, which will be defined below. It is a powerful tool for solving problems, especially ones involving many circles. Using inversion, some of the circles may transform to lines, and roughly speaking, we can say that inversion transforms problems for circles to problems for lines, which are often much easier to solve.

Definition 1. Let O be a point in the plane and r be a positive number. An *inversion* with center O and coefficient r is a transformation that maps every point $M \neq O$ onto the point $M' \in OM^{\rightarrow}$ such that $OM \cdot OM' = r^2$ (Figure 1.1). The circle ω with center O and radius r is called the *circle of inversion*. Points M and M' are called *inverse to each other*.

The properties of inversions are studied in many geometrical books, for example, Bankov and Vitanov (2003), Johnson (1960), etc. The most

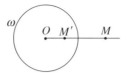

Figure 1.1. Definition of inversion.

important of these are the following:

 (i) An inversion is a one-to-one correspondence between the set of all points in the plane (without the center of the inversion) and itself.

 (ii) If a point lies on the circle of inversion, it is inverse to itself.

(iii) If a point is outside the circle of inversion, its inverse point is inside the circle of inversion, and vice versa.

This last property is used for hunting a lion in the joke at the beginning of this chapter. However, the hunter must be careful in at least two important ways. First, not only will the lion get into the interior of the cage, but also will all other things that are initially on its outside. Second, if the hunter is close to the center of the inversion, he will end up so far away from the cage that he may never find it. The latter follows from the definition: the product $OM \cdot OM' = r^2$ is a constant, and if one of the factors is very small (i.e. close to 0), the other one will be very large.

(iv) If a point M is outside the circle of inversion, its inverse point is the midpoint of the chord formed by the tangent points of the tangent lines from M to the circle of inversion (Figure 1.2).

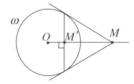

Figure 1.2. Image of a point.

 (v) Let the points M, N be different from O and points O, M, N be non-collinear. If the inverse points of M and N are M' and N', respectively, then triangles OMN and $ON'M'$ are similar (Figure 1.3) with $\angle OMN = \angle ON'M'$ and $\angle ONM = \angle OM'N'$.

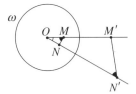

Figure 1.3. Similar triangles.

The so-called inversion distance formula then follows from property (v), namely $M'N' = \frac{r^2}{OM \cdot ON} \cdot MN$ or equivalently $MN = \frac{r^2}{OM' \cdot ON'} \cdot M'N'$.

(vi) A line passing through the center of the inversion is transformed onto itself.

(vii) A circle not passing through the center of the inversion is transformed onto a circle also not passing through the center of the inversion. The circle and its image are homothetic with the center of the inversion as the center of the homothety.

(viii) A line not passing through the center of the inversion is transformed to a circle passing through the center of the inversion, and vice versa, a circle passing through the center of the inversion is transformed to a line not passing through the center of the inversion.

The last property can be used to transform problems for circles to problems for lines.

It is easy to see that if k is a circle passing through the center of the inversion and l is its inverse image (see property (viii) above), the interior points of k are transformed to one side of l, and the exterior points of k to the other side of l.

(ix) For any two circles without a common point, there is an inversion that transforms these circles onto two concentric circles.

2. Lines and Circles

Lines and circles are different figures in Euclidean geometry with different geometrical properties. In some contexts however, because of property (viii) lines and circles can be considered alike. Some examples are presented in the following.

The first problem is from the Federal Mathematics Competition (*Bundeswettbewerb Mathematik*) in Germany, 1980.

Problem 2.1. Let M be a set of $2n + 3$ points in the plane (with n a positive integer), such that no three of which lie on a straight line and no four of which lie on a circle. Prove that there is a circle passing through three of the points of M that contains exactly n of the other points in its interior (with the other n outside the circle).

Solution. Let A and B be points of M, such that all the other $2n + 1$ points of M lie in the same semi-plane bounded by the line AB. Since no four of the points of M lie on a circle, all angles $\angle AXB$, where $X \in M \backslash \{A, B\}$, are different. Label points of $P_1, P_2, P_3 \ldots, P_{2n+1}$ of $M/[A, B]$ in such a way that $\angle AP_1B < \angle AP_2B < \angle AP_3B < \ldots < \angle AP_{2n+1}B$ holds. The circumcircle of triangle A, B, P_{n+1} then has the required property, since it contains the points $P_{n+2}, P_{n+3}, \ldots, P_{2n+1}$ in its interior, but points $P_1, P_2, \ldots, P_n$ lie outside the circle.

The next problem looks similar but considers a line instead of a circle.

Problem 2.2. Let M be a set of $2n + 2$ points on the plane (with n a positive integer) such that no three of them lie on a straight line. Prove that there is a line passing through two points of M, such that exactly n of the other points lie on each side of the line.

Solution. Let p be a line passing through a point A of M, such that all the other $2n+1$ points of M lie in the same semi-plane bounded by p. Start rotating p about A in a counter-clockwise direction. Since no three of the points of M lie on a line, the rotating line p passes consecutively through the points $P_1, P_2, P_3, \ldots, P_{2n+1}$ of $M \backslash \{A\}$. The line passing through the points A and P_{n+1} has the required property, because the points $P_{n+2}, P_{n+3}, \ldots, P_{2n+1}$ are on one side of this line, but the points $P_1, P_2, \ldots, P_n$ are on the other side.

These two problems are equivalent. More interesting than simply solving the problem is to show that the validity of the claim in Problem 2.1 follows directly from the validity of the claim in Problem 2.2. To see this, let M be a set of $2n + 3$ points in the plane such that no three of them lie on a straight line and no four of them lie on a circle. Choose a point O in M and consider the inversion with center O and an arbitrary radius r. Let N denote the set

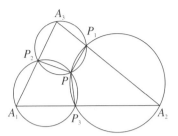

Figure 2.1. Miquel's theorem.

of inverse images of the remaining $2n + 2$ points of M. Since no three of them lie on a line, according to the statement of Problem 2.2, there is a line p passing through two of the points of N, such that exactly n of the other points lie on one side of the line (and the other n on the other side). The inverse of p is then a circle c with the required property.

The same method is used to transform Miquel's theorem into the so-called Six Circle theorem of Miquel (Bankov and Vitanov, 2003). Theorems 2.1 and 2.2 are among a set of wonderful classical theorems in plane geometry named after Auguste Miquel, a French mathematician from the 19th century. Only one of them can actually be found in Miquel's publications (Miquel, 1838), namely the following.

Theorem 2.1 (Miquel). *Let $A_1 A_2 A_3$ be a triangle, with points P_1, P_2, and P_3 on the lines $A_2 A_3$, $A_1 A_3$, and $A_1 A_2$, respectively (Figure 2.1). The three circumcircles of triangles $A_1 P_2 P_3$, $P_1 A_2 P_3$, and $P_1 P_2 A_3$ then intersect in a single point P (called the Miquel point).*

Proof. Let P be the common point of the circumcircles of triangles $A_1 P_2 P_3$ and $P_1 P_2 A_3$ that is different from P_2. Then $\angle P P_2 A_1 = 180° - \angle P P_3 A_1$ or $\angle P P_2 A_1 = \angle P P_3 A_1$ must hold depending on the positions of the points A_1, P, P_2, and P_3 on the circle. Both equations show that the angles between the lines $A_1 A_3$ and $P P_2$ and the lines $A_1 A_2$ and $P P_3$ are equal, i.e. $\angle (A_1 A_3, P P_2) = \angle (A_1 A_2, P P_3)$. Similarly, using the other circle, we obtain $\angle (A_1 A_3, P P_2) = \angle (A_2 A_3, P P_1)$. We therefore have $\angle (A_2 A_3, P P_1) = \angle (A_1 A_2, P P_3)$. It follows that either $\angle P P_1 A_2 = 180° - \angle P P_3 A_2$ or $\angle P P_1 A_2 = \angle P P_3 A_2$ holds. Either of these equations implies that P must lie on the circumcircle of triangle $P_1 A_2 P_3$. □

Theorem 2.2 (Miquel's Six Circle Theorem). *Four points, A, B, C, and D are given on a circle o. Four other circles k, l, m, and n pass through each adjacent pair of these points. The alternate intersections of these four circles at E, F, G and H, respectively, then lie on a common circle (or on a common line) (Figure 2.2).*

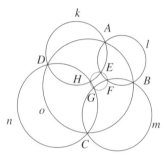

Figure 2.2. Miquel's six circle theorem.

These two theorems are equivalent. Here is an argument to show that Theorem 2.2 follows directly from Theorem 2.1. Consider the inversion with center A and an arbitrary coefficient r. Let the inverse of each figure be denoted by the same letter as the original with an added index 1. From the property (viii) of inversion, it follows that k_1 (i.e. the image of k) is a line containing the points D_1, H_1, and E_1 (Figure 2.3), l_1 is a line containing the points B_1, F_1, and E_1, and o_1 is a line containing the points D_1, C_1, and B_1. Also, m_1 is a circle containing the points B_1, F_1, G_1, and C_1 and n_1 is a circle containing the points D_1, H_1, G_1, and C_1. According to Theorem

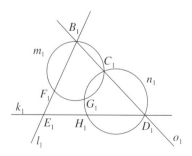

Figure 2.3. Inversion of Figure 2.2.

2.1, points E_1, F_1, G_1, and H_1 lie on a common circle. Therefore, points E, F, G and H also lie on a common circle (or on a common line).

The above reasoning shows how a suitable inversion may transform a problem concerning "many" circles to a problem that contains "less circles but more lines". Since problems for lines are generally easier to solve, inversion yields a path to an easier solution of the original problem. Amazing, isn't it?

3. Two Famous Results from Plane Geometry

Miquel's theorems are classical results from plane geometry. Here are two other examples of beautiful geometrical theorems connected to the names of well-known mathematicians. The reason to bring them up in this context is that inversion can help to prove them as well.

The first example is connected with the figure known as the *arbelos*. This figure consists of three collinear points A, B, and C, together with three semicircles with diameters AB, AC, and BC, as shown in Figure 3.1. It was named after a shoemaker's knife because its shape resembles it. Many interesting properties of the arbelos have been studied by mathematicians (i.e. Bankoff, 1974; Cadwell, 1966; Hood, 1961). Here, we consider a statement believed to have been proven by Pappus.

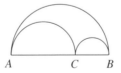

Figure 3.1.   Arbelos.

Theorem (Pappus). *Consider a chain of circles* $c_1, c_2, \ldots, c_n, \ldots$ *inscribed in an arbelos as shown in Figure* 3.2. *(Circle c_1 is tangent to the three semicircles of the arbelos, and for $n > 1$, c_n is tangent to two of the semicircles of the arbelos and the circle c_{n-1}.) For any $n = 1, 2, \ldots$, let r_n denote the radius of c_n and y_n the distance from the center of c_n to the baseline AB. Then $y_n = 2nr_n$.*

The proof of this theorem is relatively easy if we apply the inversion with center A that transforms B to C (i.e. the radius of the inversion is

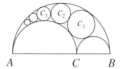

Figure 3.2. Pappus' theorem.

$r = \sqrt{AB \cdot AC}$). This inversion transforms the semicircle with diameter AB into a ray with origin C and perpendicular to AB. Simultaneously, it transforms the semicircle with diameter AC into a ray with origin B and also perpendicular to AB. Furthermore, it transforms the semicircle with diameter BC onto itself. The chain $c_1, c_2, \ldots, c_n, \ldots$ is therefore transformed onto a chain $c'_1, c'_2, \ldots, c'_n, \ldots$ of touching equal circles, each of which is tangent to the two rays as shown on Figure 3.3. If r'_n is the radius of c'_n and y'_n is the distance from the center of c'_n to the baseline AB, it is clear that $y'_n = 2nr'_n$. We can now use property (vii) of the inversion, which states that the circle and its inverse image are homothetic with the center of homothety lying in the center of inversion used to obtain the result.

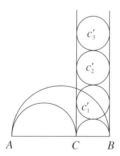

Figure 3.3. Proof of Pappus' theorem.

Pappus of Alexandria was one of the last great Greek mathematicians of antiquity. He lived in the fourth century. This means that Pappus was not able to make use of this method, since inversion was not discovered for another 15 centuries after Pappus. In 1981, Bankoff (1981) wrote an interesting essay on how Pappus could have proven this theorem.

The second example is connected with the Swiss mathematician of the 19th century, Jakob Steiner. He considered two circles k_1 and k_2, with k_2

in the interior of k_1. In the area between the two circles, we have a chain of touching circles $c_1, c_2, \ldots, c_n$ such that each circle is tangent to k_1, k_2, and both of its neighbours (Figure 3.4). For some circles k_1 and k_2, it is possible to find an n such that c_n is tangent to k_1, k_2, c_{n-1}, and c_1, and in some cases this is not possible. This depends on the radii of k_1, k_2 and their mutual position.

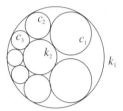

Figure 3.4. Steiner's theorem.

Theorem (Steiner). *In the above configuration, let k_1 and k_2 be circles for which there exists a chain $c_1, c_2, \ldots, c_n$ such that c_n is tangent to k_1, k_2, c_{n-1}, and c_1. Then there exist infinitely many chains of this type. In other words, for any circle t_1 tangent to k_1 and k_2, there exists a chain of touching circles $t_1, t_2, \ldots, t_n$ such that each circle is tangent to k_1, k_2, and both of its neighbors, and t_n is tangent to k_1, k_2, t_{n-1}, and t_n.*

Steiner was more fortunate than Pappus because the concept of inversion was well known by his time, and the theorem is easy to prove with this tool. Simply apply property (ix) to find an inversion that transforms k_1 and k_2 onto two concentric circles k_1', k_2'. For k_1', k_2' the result is obvious. Using the inverse of this inversion then yields the required result.

The question that remains is under which conditions such a chain of touching circles $c_1, c_2, \ldots, c_n$ with c_n tangent to k_1, k_2, c_{n-1}, and c_1 exists. We will not be discussing this here, but the answer can be found in Prassolov (2006) among other places.

4. Inversion in Competition Problems

This section contains some examples of competition problems that can be solved using inversion. One reason to present these here is to show that the often heard claim that "geometry is becoming less popular in school mathematics and in competitions" is really only a myth.

The beauty of these competition problems is that they do not directly refer to inversion in the statement of the problem. There is not much creativity involved in solving a problem of the type: "A geometrical construction and its image under a certain inversion is given. Prove such and such." In these problems, the required inversion is not given explicitly. Discovering the appropriate inversion and applying it is an intrinsic part of the solution. This requires a great deal of creativity. Such problems are perfect examples of the beauty of geometry.

Only two examples are presented in this chapter. Many others can be found in Chapter 8 of Evan (2016).

Problem 4.1 (IMO, 1996). Let P be a point inside $\triangle ABC$ such that $\angle APB - \angle C = \angle APC - \angle B$. Let D and E be the incenters of $\triangle APB$ and $\triangle APC$, respectively. Show that AP, BD, and CE meet at a point.

Solution. Apply an inversion with center A and an arbitrary radius r. Let B', C', and P' be the images of B, C, and P, respectively. Because of property (v) of inversion, there are four pairs of equal angles marked with one and the same sign in Figure 4.1. The given equation becomes $\angle B'C'P' = \angle C'B'P'$, i.e. $P'B' = P'C'$. From the inversion distance formula, we obtain $P'B' = \frac{r^2}{AP \cdot AB} \cdot PB$ and $P'C' = \frac{r^2}{AP \cdot AC} \cdot PC$. The equation $P'B' = P'C'$ is then equivalent to $\frac{BA}{BP} = \frac{CA}{CP}$. Let the angle bisector BD in $\triangle APB$ meet AP in K. Then we have $\frac{AK}{KP} = \frac{BA}{BP} = \frac{CA}{CP}$. It therefore follows that the angle bisector CE in $\triangle APC$ meets AP in the same point K. □

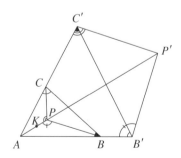

Figure 4.1. Solution to Problem 4.1.

Problem 4.2 (International Mathematics Tournament of Towns, 2016–2017). Quadrilateral $ABCD$ is inscribed in a circle k, whose center O is not on either of the diagonals of the quadrilateral. The midpoint of the diagonal BD lies on the circumcircle k_1 of triangle AOC. Prove that the midpoint of the diagonal AC lies on the circumcircle k_2 of triangle BOD.

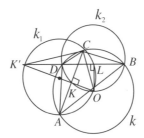

Figure 4.2. Solution to Problem 4.2.

Solution. Let K and L denote the midpoints of AC and BD, respectively (Figure 4.2). Then we have $OK \perp AC$ and $OL \perp BD$. Let K' be a point on k_1 such that OK' is a diameter of k_1. Then $\angle OLK' = 90°$, which implies $K' \in BD$. Also, we have $\angle OCK' = 90°$ and therefore $OK \cdot OK' = OC^2 = r^2$, where r is the radius of k. The last equation gives us a hint for introducing an inversion. From this equation, it follows that the inversion with center O and radius r transforms K onto K' and the line AC onto the circle k_1. Similarly, we also obtain that the same inversion transforms the line BD onto the circle k_2. Since $K' \in BD$, its image K lies on k_2. □

5. Problems for the Reader

Problem 5.1. Four circles k_1, k_2, k_3, and k_4 are given in such a way that k_1 is externally tangent to k_2 and k_4 in points A and D, respectively, and k_3 is externally tangent to k_2 and k_4 in points B and C, respectively. Prove that points A, B, C, and D lie on a common circle.

Hint. Consider an inversion with center A and an arbitrary coefficient. The problem is then transformed to the following easier problem: Let p and q be parallel lines, and A and C points on p and q, respectively. Let k_1

and k_2 be circles externally tangent in a point E, such that k_1 is tangent to p in A and k_2 is tangent to q in C. Prove that A, E, and C lie on a common line.

Problem 5.2 (Russian Mathematical Olympiad, 1995). A semicircle with diameter AB and center O is given. A line intersects the semicircle at C and D, and line AB at M ($MB < MA$, $MD < MC$). Let K be the second point of intersection of the circumcircles of triangles AOC and DOB. Prove that $\angle MKO = 90°$.

Hint. Consider the inversion with center O and radius $r = OA = OB$. By property (ii) of inversion, each of the points A, B, C, and D is inverse to itself. The image of the circle through A, O, and C is the line AC and the image of the circle through D, O, and B is the line DB. Hence, the inverse of K is the intersection K_1 of lines AC and DB. Also, the inverse of M is the intersection M_1 of line AB with the circumcircle of OCD. Then the inverse of the line MK is the circumcircle of OM_1K_1. According to property (v) of inversion, $\angle MKO = 90°$ holds if and only if $\angle K_1M_1O = 90°$. To prove this last equation, show that the circumcircle of OCD is the nine-point circle of ABK_1. (The nine-point circle of a triangle is the circle that contains the feet of the altitudes of the triangle and the midpoints of its sides.)

Bibliography

Bankoff, L. (1974). Are the twin circles of archimedes really twins? *Mathematics Magazine* 47(4), 214–218.

Bankoff, L. (1981). How did Pappus do it? In *The Mathematical Gardner*, edited by David Klarner, Wadsworth International, Belmont, California, pp. 112–118.

Bankov, K., and Vitanov, T. (2003). *Geometry*, Anubis (in Bulgarian).

Cadwell, J. (1966). *Topics in Recreational Mathematics*, Cambridge University Press, Cambridge, pp. 44–45.

Evan, C. (2016). *Euclidean Geometry in Mathematical Olympiads*. The Mathematical Association of America, USA.

Hood, R. (1961). A chain of circles, *The Mathematics Teacher* 54(3), 134–137.

Johnson, R. (1960). *Advanced Euclidean Geometry*, New York: Dover Publications.

Miquel, A. (1838). Théorèmes de Géométrie, *Journal de Mathématiques Pures et Appliquées* 1, 485–487.

Petard, H. (1938). A contribution to the mathematical theory of big game hunting, *The Mathematical Monthly* 45(7), 446–447.

Prassolov, V. (2006). *Problems on 2D Geometry* [Chapter 28, Problem 28.41], Moscow: MCNMO, pp. 523–524 (in Russian).

Chapter 1.4

Sangaku: Traditional Japanese Mathematics

Hidetoshi Fukagawa

Kani, Japan

1. Introduction

One of the major epochs of Japanese history, the Edo period (1600–1868), was distinguished by the fact that Japan enjoyed freedom from warfare at home and abroad for more than two centuries. As a consequence of the

Tokugawa family's adoption of the policy of *National Seclusion* in order for Japan not to be conquered by colonial European powers like Portugal or Spain, unique traditions arose all over Japan. Ordinary people, samurai and farmers alike, would form groups in their own neighborhoods or in nearby private schools, where people learned "reading, writing and arithmetic". In school arithmetic, students learned to work with the traditional calculators known as *soroban,* and many talented students studied mathematics. Some people who wanted to display their abilities to others would hang the wooden tablets called *sangaku*, on which geometric problems were drawn in beautiful colors, near a temple or shrine. Despite the fact that there was no formal academia in the Edo-period, the traditional Japanese mathematics known as *wasan* flourished.

After the Edo-age, the new government abandoned traditional Japanese mathematics and introduced western mathematics. In every school, teachers taught western mathematics, but some who loved the traditional Japanese mathematics continued to hang sangaku. In the beginning of the 20th century, the tradition of hanging sangaku ceased. We now know of one thousand tablets that have survived.

Let us take a look at some of the sangaku produced at certain private schools during the Edo-period.

2. Sangaku Problems of Hikuma Shrine (1797)

These tablets were originally hung in 1797. For many years, no one took any interest in these sangaku, and they were assumed to have disappeared. In 1994, Mr. Kinji Hatano was searching for forgotten sangaku and discovered them in an abandoned room of the small shrine "Hikuma-Jinzya". The discovery of these tablets was a great achievement for Mr. Hatano.

This particular sangaku consists of three tablets. In total, they cover an area 520 cm wide and 34 cm high. Thirty problems were drawn on it, all of which are well suited for young students. Perhaps, you would like to use these problems in your competitions.

Problem 2.1. We want to measure the distance AX, where A is a point on one side of a river and X a tree on the other side. In order to do this, we have a rectangular board $ABCD$ and a measuring tape at our disposal. We place the board in such a way that we see the point X on the extended line AD. We then translate the board to $A'B'C'D'$ in such a way that we see the point X on the extended line $D'B'$ and AA' is perpendicular to $D'B'$. We then draw a right triangle $A'PB'$ on the board. We can then show that it is possible to determine the distance AX by virtue of $\frac{AX}{AP} = \frac{A'B'}{A'P}$ or $AX = \frac{A'B'}{A'P} \cdot AP$.

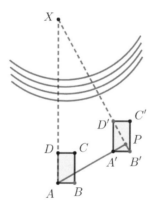

This problem was proposed by Syouichi Ishiguro who was a student of the mathematician Touko Watanabe.

Problem 2.2. A square with sides of length 12 is inscribed in an isosceles triangle ABC and three touching circles of diameters a, b and b are given as shown. If $ab = 32$, determine the value of a.

Answer. $a^3 - 112a + 384 = 0$ or $a = 8$.

This problem was proposed by Isomura Chikayuki, who was a student of the Saitou School.

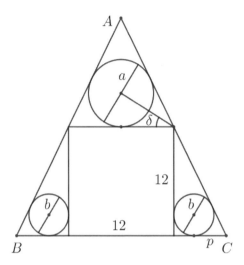

Solution by H. Fukagawa. From the figure, we have $\tan \delta = \frac{a/2}{6} = \frac{a}{12} = \frac{b/2}{p} = \frac{b}{2p}$ and $\tan 2\vartheta = \frac{12}{p+b/2} = \frac{24}{2p+b} = \frac{2\tan\vartheta}{1-\tan^2\vartheta}$. These relations give us $p = \frac{6b}{a}$ and $\frac{24}{6b/a+b} = \frac{2(a/12)}{1-(a/12)^2}$ or $\frac{1}{12b+ab} = \frac{1}{144-a^2}$. Substituting $b = \frac{32}{a}$, we obtain $a^3 - 112a + 384 = 0$, which can be factorized as $a^3 - 112a + 384 = (a+12)(a-8)(a-4) = 0$, and the solutions of this equation are $a = 8, 4$ and -12.

The result is written in the form $a = \sqrt{6^2 - 32} + 6 = 8$ on the tablet.

Problem 2.3. Two adjoining small squares of equal size and a circle are inscribed in a square with sides of length 169 as shown in the figure. The diameter $2r$ of the small circle is equal to the lengths of the sides of the two small squares. Determine the value of $2r$.

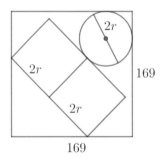

Answer. $2r = 74.5217$.

This problem was proposed by Ishizuka Masakatsu.

Solution by H. Fukagawa. In the diagonal of the large square, we have

$$2r + 2r + r + \sqrt{2}r = 169\sqrt{2}.$$

From this, we obtain $r = \frac{169\sqrt{2}}{5+\sqrt{2}} = \frac{169(5-\sqrt{2})\sqrt{2}}{23}$ or $2r = 74.52264871$.

Problem 2.4. We are given a trapezoid $ABCD$ with $AD = BC = CD = 13$, $AB\|CD$, and $AB = x$. The area S of the trapezoid is equal to 96. Determine the value of $AB = x$ and the height h of the trapezoid.

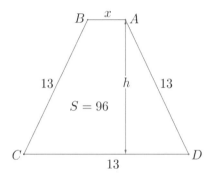

Answer. $AB = x = 3$ and $h = 12$.

This problem was proposed by Watanabe Touko.

Solution by H. Fukagawa. We have $S = 96 = \frac{13+x}{2} \cdot h$, and therefore $h = \frac{192}{13+x}$, and the Pythagorean theorem gives us $13^2 = h^2 + \left(\frac{13-x}{2}\right)^2$, or $169 = \left(\frac{192}{13+x}\right)^2 + \left(\frac{13-x}{2}\right)^2$. Simplifying yields the equation

$$x^4 - 1014x^2 - 17576x + 61773 = (x-3)(x^3 + 3x^2 - 1000x - 205891) = 0,$$

and one of the roots of this equation (the only integer root) is $x = 3$. It is then easy to calculate $h = \frac{192}{13+3} = 12$.

The fourth degree equation is given on the tablet.

Problem 2.5. Two roads, each of which is 2 m wide, divide a given square with sides of length 169 m into three sections of equal area S, as shown in the figure. Determine the length of x.

Answer. $x = 192.842\,m$.

This problem was proposed by Ishizuka Masakatsu.

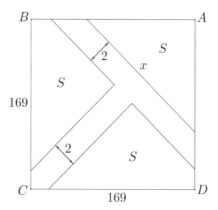

Solution by H. Fukagawa. As a first step, we determine the area covered by the roads. This area is equal to $2x + 4 + 2(169\sqrt{2} - \frac{x}{2} - 2) - 1 = x - 1 + 338\sqrt{2}$, and it therefore follows that $S = \frac{1}{3}(169^2 - x + 1 - 338\sqrt{2})$. We therefore obtain the equation

$$S = \frac{x^2}{4} = \frac{1}{3}(169^2 - x + 1 - 338\sqrt{2}).$$

From this, we calculate $3x^2 = 4(169^2 - x + 1 - 338\sqrt{2})$, or $3x^2 + 4x = 4(169^2 - 338\sqrt{2})$, and therefore $\left(x + \frac{2}{3}\right)^2 = \frac{4}{9}(3(169^2 - 338\sqrt{2} + 1))$. This then yields the value

$$x = \frac{2}{3}\left(\sqrt{3(169^2 - 338\sqrt{2}) + 1} - 1\right) \approx 192.839.$$

Problem 2.6. We are given a rhombus $ABCD$ inscribed as shown in a semicircle, the diameter of which is AE. A circle of diameter p is inscribed in the rhombus and q is the diameter of the circle touching the sides CD and DE and the semicircle. If we are given $p = 10$, determine q.

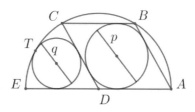

Answer. $q = 7.698$.

This problem was proposed by Asawaka Nobusada.

Solution by H. Fukagawa. As a first step, we must prove that D is the midpoint of the semicircle. The line BD is perpendicular to the chord AC, which implies that the line BD passes through the midpoint of the semicircle, and since AE is its diameter, it follows that D is the midpoint of the semicircle. The sides AB and BC are therefore sides of the inscribed hexagon in the circle with midpoint D passing through A. This implies both $BD = \frac{4}{\sqrt{3}}p$ and $DT = 3q$, and we therefore obtain $q = \frac{4}{3\sqrt{3}}p \approx 7.698$.

Problem 2.7. We are given an equilateral triangle ABC with sides of length a. Three lines AA', BB' and CC' are drawn from the vertices A, B and C such that the interior triangle $A'B'C'$ is also equilateral and the inradius r of $A'B'C'$ is the same as the inradii r of the triangles $AB'B$, $BC'C$ and $AA'C$. Determine r in terms of a.

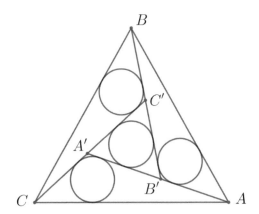

Answer. $r = \frac{\sqrt{7}-\sqrt{3}}{8} \cdot a \approx 1.142125 \cdot a$.

This problem was proposed by Shirai Tametsuna.

Solution by H. Fukagawa. Let P denote the point of tangency of the incircle of $AA'C$ with the line $A'C$ and Q its point of tangency with the line AA'.

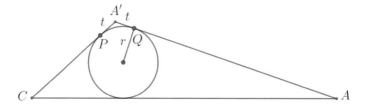

Since $\angle AA'C = 120°$, we then have $t = A'P = \frac{r}{\sqrt{3}}$. For $A'C = p$ and $AA' = q$, we have $a = CA = CP + AQ = p - t + q - t$ and therefore $p + q = a + \frac{2}{\sqrt{3}\cdot r}$. We now consider the areas of the triangles. From

$$A(\triangle CAA') = \frac{1}{2}r \cdot (a + p + q) = r\left(a + \frac{1}{\sqrt{3}\cdot r}\right),$$

we obtain

$$A(\triangle ABC) = \left(\frac{\sqrt{3}}{4}\right)a^2 = 3 \cdot A(\triangle CAA') + A(\triangle A'B'C')$$

$$= 3r\left(a + \frac{1}{\sqrt{3r}}r\right) + 3\sqrt{3}r^2,$$

and this equation yields the answer $r = \frac{\sqrt{7}-\sqrt{3}}{8} \cdot a$.

Problem 2.8. We are given a square with sides of length $a = 169$. Four lines are drawn as shown in the interior of the square, two from the vertex B and two from the vertex D, such that they are tangent to three small circles of equal size and two pairs are each parallel. Two circles are each tangent to two sides of the square and one circle is tangent to all four lines. Determine the common diameter $d = 2r$ of the circles.

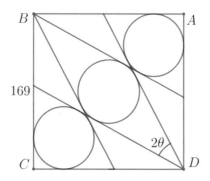

Answer. $d = 2r = 68.27$.

This problem was proposed by Setaka Saitou.

Solution by H. Fukagawa. Defining the angle θ as shown in the figure, we have $\sin\theta = \frac{r}{\frac{a}{\sqrt{2}}} = \frac{\sqrt{2}r}{a}$ and $\sqrt{2}a = AC = 2\sqrt{2}r + \frac{4r}{\cos\theta}$. This then yields $\sqrt{2}(a - 2r) = \frac{4r}{\cos\theta} = \frac{4ar}{\sqrt{a^2 - 2r^2}}$. This is equivalent to

$$(a - 2r)^2(a^2 - 2r^2) = 8a^2r^2(a^2 - 4ar + 4r^2)(a^2 - 2r^2) = 4a^2r^2,$$

and finally, we get $a^4 - 2a^2r^2 - 4a^3r + 8ar^3 + 4a^2r^2 - 8r^4 = 4a^2r^2$, or

$$8r^4 - 8ar^3 + 2a^2r^2 + 4a^3r - a^4 = 0.$$

I was not able to obtain the result on the tablet. I hope that someone will be able to derive this solution.

Problem 2.9. Let G denote the centroid of triangle ABC with $AG = 17$, $BG = 20$, and $CG = 15$. Determine the length of the side BC.

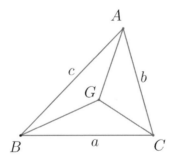

Answer. $BC = 31$.

This problem was proposed by Nagasaka Tsunenori.

Solution by H. Fukagawa. If we let M denote the midpoint of BC, we have

$$AM = \frac{1}{2}\sqrt{2(b^2 + c^2) - c^2}$$

by Apollonius' theorem, and therefore $AG = p = \frac{1}{3}\sqrt{2(b^2 + c^2) - a^2}$. Similarly, we also have $BG = q = \frac{1}{3}\sqrt{2(a^2 + c^2) - b^2}$ and

$CG = r = \frac{1}{3}\sqrt{2(a^2 + b^2) - c^2}$. Adding the relationships

$$9p^2 = 2(a^2 + c^2) - a^2, \ 9q^2 = 2(a^2 + c^2) - b^2, \ \text{and} \ 9r^2 = 2(a^2 + b^2) - c^2$$

gives us $3(p^2 + q^2 + r^2) = a^2 + b^2 + c^2$, and substituting in the first relationship then yields $9p^2 = 2(3p^2 + 3q^2 + 3r^2 - a^2) - a^2$, and therefore $3a^2 = 6(q^2 + r^2) - 3p^2$. From this, we obtain $a^2 = 2(q^2 + r^2) - p^2 = 2(20^2 + 15^2) - 17^2 = 961$ or $a = \sqrt{961} = 31$.

Problem 2.10. In a semicircle with diameter $AB = 2R$, the two chords AP and BQ are each tangent to two small circles of equal size as shown in the figure. If $AP = BQ = 10$, determine the length of $AB = 2R$.

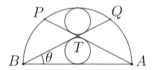

Answer. $2R \approx 11.18$.

This problem was proposed by Isomura Chikayuki.

Solution by H. Fukagawa. Let H denote the midpoint of AB and θ as shown in the figure, we have $\cos\theta = \frac{BQ}{AB} = \frac{BH}{BT}$ with $AB = 2R$, $BQ = 10$, and $BT = \sqrt{R^2 + \frac{R^2}{4}} = \frac{\sqrt{5}}{2}R$. Since $BH = R$ holds, we therefore obtain

$$AB = \frac{BQ \cdot BT}{BH} = \frac{10\left(\frac{\sqrt{5}}{2}\right)R}{R} = \frac{10\sqrt{5}}{2} = 5\sqrt{5} \approx 11.18.$$

Problem 2.11. At a party, n guests and m attendants participate and drink many cups of sake. Each guest drinks sake from a 350-ml mug and each attendant drinks from a 250-ml mug. Altogether, they empty one entire 3.5 liter barrel of sake. The total amount of sake consumed by the guests is equal to the total amount consumed by the attendants. Determine the number of guests and attendants at the party.

Answer. $n = 5$ and $m = 7$.

This problem was proposed by Isomura Chikayuki.

Solution by H. Fukagawa. We are given $350n = 250m$, or $7n = 5m$, and $350n + 250m = 3500$. We therefore obtain $m = 7$, $n = 5$.

Problem 2.12. We are given a trapezoid $ABCD$ with $BC\|AD$, and $AD\perp DC$. Three roads AE, FG, and HC are constructed such that the triangles $\triangle DAE$, $\triangle EFG$ and $\triangle GHC$ are similar. Furthermore, we are given $BC = b = 10\,m$, $AD = a = 80\,m$ and $CD = h = 105\,m$. Determine the ratio k of the lengths of the legs of the similar triangles.

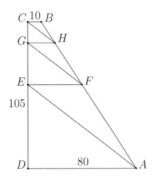

Answer. The ratio is equal to $k = \frac{3}{4}$.

This problem was proposed by Ishizuka Masakatsu.

Solution by H. Fukagawa. From $b = 10$, $GH = 10r$, $EF = 10r^2$, $AD = 10r^3 = 80$, we obtain $r = 2$. From this, we have $CG = c$, $GE = 2c$, $ED = 4c$, and $CD = 7c = 105$, and therefore $c = \frac{105}{7} = 15$. We therefore obtain $k = \frac{c}{GH} = \frac{15}{20} = \frac{3}{4}$.

Problem 2.13. We are given a right triangle ABC. Its incircle has the midpoint R and radius r. Furthermore, we are given a rhombus $ADEF$ with $AD = DE = EF = FA$. A circle with radius t and the midpoint T is tangent to the circle of radius r and three sides of the rhombus. We are given $2t = 1$. Determine $2r$.

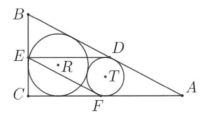

Answer. On the tablet, the third degree equation $r^3 - 6tr^2 + 17t^2r - 16t^3 = 0$ and the answer $2r = 1.61173$ are given.

This problem was proposed by Ishizuka Masakazu.

Solution by H. Fukagawa. The midpoints R and T both lie on the angle bisector AE. Therefore, we have $\frac{CE}{AC} = \frac{r}{AC-r} = \frac{t}{AC-r-2\sqrt{rt}}$, and define k by $\frac{2t}{b} = \frac{r}{b-r} = \frac{t}{b-r-2\sqrt{rt}} = \frac{1}{k}$. We now have the equations

$$b = 2kt, \tag{1}$$

$$b - r = rk, \tag{2}$$

$$b - r - 2\sqrt{rt} = tk. \tag{3}$$

From (1), we eliminate b and so (2) yields $k(2t - r) = r$ or $k = \frac{r}{2t-r}$. From (3), we then obtain $r + 2\sqrt{rt} = \frac{tr}{2t-r}$, which is equivalent to $(\sqrt{r} + 2\sqrt{t})(2t - r) = t\sqrt{r}$ or $\sqrt{r}(r - t) = 2\sqrt{t}(2t - r)$. Squaring both sides gives us $r(r - t)^2 = 4t(2t - r)^2$ or $r^3 - 6tr^2 + 17t^2r - 16t^3 = 0$, and the real solution of this is $r \approx 1.6117086$ if $t = 1$, as stated on the tablet.

The author would like to express many thanks to Prof. Yoshiyuki Kitaoka for this calculation.

Problem 2.14. A square is inscribed in a right triangle ABC as shown in the figure, such that the vertex P of the square lies on the hypotenuse AB of the triangle and two sides lie on AB and BC, respectively. The altitude CH is perpendicular to AB. Determine BH in terms of $p = PH$ and $h = CH$.

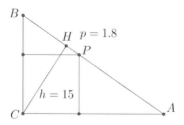

Answer. $BH = \sqrt{\left(\frac{h}{2}\right)^2 - p(p + h)} + \frac{h}{2}$. If $h = 15$ and $p = 1.8$, then $HB = 12.6$. This problem was proposed by Yamamoto Nobuyuki.

The author has not yet been able to solve this problem.

Problem 2.15. The three points O, P and Q are the midpoints of the incircles of the right triangles ABC, BHC and ACH, respectively, where H is the foot on the side AB from the vertex C. Determine the inradius r of ABC in terms of $p = PO$ and $q = OQ$.

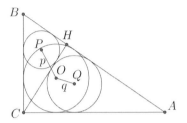

Answer. $p = 1.007$ and $q = 1.0198$, then $2r = 2$. This problem was proposed by Koike Kiyonaga. The author has not yet been able to solve this problem.

Problem 2.16. We are given a rectangle $ABCD$, $AB < AD$, and an inscribed circle with radius $r = CD/2$ tangent to the three sides BC, CD and AD. The tangent AE of this circle is drawn from the vertex A, as is the incircle of triangle ABE, with radius t. Show that $t = AD - AE$ holds.

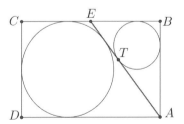

Answer. If $AD = a = 6$ and $AE = k = 5$, then $t = a - k = 1$.
This problem was proposed by Suzuki Naokata.

Solution by H. Fukagawa. If we let k denote the length of AE and p the tangent distance from E to the large circle, we obtain

$$k = AE = AT + ET = (b - t) + (EB - t) = b - 2t + \left(a - \frac{b}{2} - p\right)$$

$$= a + \frac{b}{2} - p - 2t.$$

Substituting $p = k - \left(a - \frac{b}{2}\right)$ in the above equation yields $k = a + \frac{b}{2} - k + a - \frac{b}{2} - 2t$ or $2t = 2a - 2k = 2(a - k)$. This is a very nice problem for students.

Problem 2.17. Three chains of five circles with radii c, b, a, b, c ($a > b > c$) as shown are all internally tangent to a large circle of radius r and to the sides of an inscribed equilateral triangle. Determine r in terms of c.

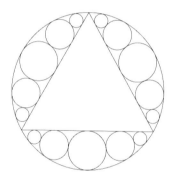

Answer. $r = \frac{100}{9}c$. If $c = 72$, then $r = 800$. This problem was proposed by Nakamura Nobumasa.

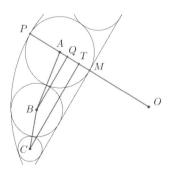

Solution by H. Fukagawa. From the figure, we have $OM = \frac{r}{2}$, and $a = r/4$. In the right triangle OBQ, we have $(r-b)^2 = (r - 2a + b)^2 + (2\sqrt{ab})^2$ or $(r-b)^2 - (\frac{r}{2} + b)^2 = 4ab = rb$, which yields $b = \frac{3}{16}r$. Similarly, in the right triangle OCT, we have

$$(r - c)^2 = \left(\frac{r}{2} + c\right)^2 + (2\sqrt{ab} + 2\sqrt{bc})^2.$$

This can be written in the form $\frac{3r^2}{4} - 3rc = \frac{3r}{4} \cdot (\sqrt{\frac{r}{4}} + \sqrt{c})^2$, and this gives us

$$9r^2 - 120rc + 400c^2 = 16cr,$$

and therefore

$$9r^2 - 136rc + 400c^2 = (9r - 100c)(r - 4c) = 0 \text{ or } r = \frac{100}{9}c.$$

Problem 2.18. A rectangle $PQRS$ in inscribed in a rhombus $ABCD$ as shown in the figure, with the vertices P, Q, R and S in the midpoints of the sides of the rhombus. Four small circles of equal radius r are inscribed as shown in the figure. We are given $PQ = 2a = 6$ and $PS = 2b = 4$. Determine the diameter $d = 2r$ of the small circles.

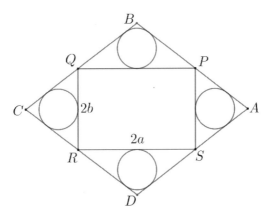

Answer. $d = 2r = 2$.
   This problem was proposed by Makino Hiroyuki.

Solution by H. Fukagawa. For $2\alpha = \angle BPQ$ and $2\beta = \angle APS$, we have $\alpha + \beta = 45°$ and therefore $1 = \tan(\alpha + \beta) = \frac{\tan\alpha + \tan\beta}{1 - \tan\alpha\tan\beta}$ for $\tan\alpha = \frac{r}{a} = \frac{r}{3}$, $\tan\beta = \frac{r}{b} = \frac{r}{2}$. Substituting gives us the equation

$$r^2 + 5r - 6 = (r-1)(r+6) = 0,$$

which yields $r = 1$ or $d = 2r = 2$.

Problem 2.19. Determine positive integers A and B such that $A - B = x^2$ and $A + B = y^3$ holds for some positive integers x and y.

Answer. $A = \frac{x^2 + y^3}{2}$, $B = \frac{y^3 - x^2}{2}$, $y > x$ with x and y either both even or both odd.

This problem was proposed by Nojima Eijiro.

Solution by H. Fukagawa. Substituting $A = \frac{x^2 + y^3}{2}$, $B = \frac{y^3 - x^2}{2}$, with x and y either both even or both odd and $y^3 > x^2$ yields $A - B = x^2$ and $A + B = y^3$. If $y = 3, x = 1$, then $A = 14$, $B = 13$, for instance. Substituting $y = 5$, $x = 3$, yields $A = 67$, $B = 58$. □

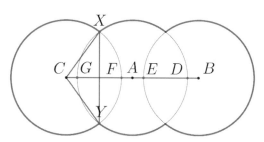

Problem 2.20. Three circles with midpoints A, B and C, each with radius $r = 5$ are given. The point A is the midpoint of BC. The circles intersect with the line BC in D, E, F and G, as shown in the figure, with $DE = 4$ and $FG = 4$. Determine the area S of the resulting shape.

Answer. $S = 191.496$.

This problem was proposed by Ikuta Hanbe.

Solution by H. Fukagawa. We can easily determine the area S of the segment CXY by

$$S = 25 \cos^{-1} \frac{3}{5} - 12 \approx 23.182 - 12 \approx 11.182.$$

The desired area is therefore $3 \times 2 \times 5 \times \pi - 4S \approx 190.89$.

Problem 2.21. The altitude AH on the side BC is drawn in triangle ABC. Determine the height $h = AH$ in terms of a, b and c.

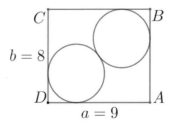

Answer. By Heron's formula, we have $h = \frac{2\sqrt{s(s-a)(s-b)(s-c)}}{a}$ for $s = \frac{a+b+c}{2}$. $AH = 16$, if $a = 42, b = 34$ and $c = 20$. This problem was proposed by Ui Sadauemon.

Solution by H. Fukagawa. Since there are two simple ways to calculate the area of a triangle, we have $A(\triangle ABC) = \frac{1}{2} ah = \sqrt{s(s-a)(s-b)(s-c)}$, and therefore $h = \frac{2\sqrt{s(s-a)(s-b)(s-c)}}{a}$ for $s = \frac{a+b+c}{2}$. For $a = 42, b = 34$, $c = 20$, this yields $s = 48$ and $h = \frac{2\sqrt{48 \times 6 \times 14 \times 28}}{42} = 16$.

Problem 2.22. Two mutually tangent circles of equal size are inscribed in the rectangle $ABCD$. Determine the radius r of the circles in terms of a and b.

Answer. $r = \frac{a+b-\sqrt{2a}}{2}$. If $a = 9$ and $b = 8$, then $r = 2.5$.
This problem was proposed by Ishiguro Syoichi.

Solution by H. Fukagawa. From the figure, we have $(2r)^2 = (a - 2r)^2 + (b - 2r)^2$ or $4r^2 - 4(a + b)r + a^2 + b^2 = 0$. This can be written as $(2r - (a + b))^2 = 2ab$, which yields $2r = a + b - \sqrt{2ab}$.

Problem 2.23. A rhombus $ABCD$ and a small circle with radius t are drawn inside a circle of radius r as shown. The circles are internally tangent in E, and BE is a diameter of the big circle. Determine r in terms of $p = \frac{AC}{2}$ and t.

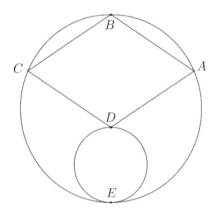

Answer. $r = \sqrt{p^2 + t^2}$. If $p = 6$ and $t = \frac{5}{2}$, then $r = \frac{13}{2}$. This problem was proposed by Nakayama Sadakatsu.

Solution by H. Fukagawa. If M is the midpoint of AC, we have $p^2 = AM^2 = BM \times ME = q(q + 2t)$ for $q = BM$. Since $2q + 2t = 2r$, we have $q = r - t$, and therefore $p^2 = (r - t)(r + t) = r^2 - t^2$ or $r = \sqrt{p^2 + t^2}$.

Problem 2.24. A circle with radius r touches two sides BA and BC of the equilateral triangle ABC and the line AD. Determine $AD = p$ in terms of $AB = BC = CA = a$ and r.

Answer. $p = \frac{(a - \sqrt{3}r)^2 + r^2}{a - 4r/\sqrt{3}}$. If $a = 10$, $r = 1$, then $p \approx 9.01869$. This problem was proposed by Ishiguro Syouichi.

Solution by H. Fukagawa. By the cosine rule, we have $p^2 = a^2 + t^2 - at$, for $BD = t$. The area of the triangle ABD is equal to $S = rs = \frac{r}{2}(a + p + t) = \frac{\sqrt{3}}{4}$ at which yields $t = \frac{2r(a + p)}{\sqrt{3}a - 2r}$. Substituting for

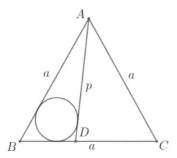

t gives us $p^2 = a^2 + \left(\frac{2r(a+p)}{\sqrt{3}a-2r}\right)^2 - a\left(\frac{2r(a+p)}{\sqrt{3}a-2r}\right)^2$ This is equivalent to $(p^2 - a^2)(\sqrt{3}a - 2r)^2 = 4r^2(a + p)^2 - 2ar(\sqrt{3}a - 2r)(a + p)$ or $(p - a)(\sqrt{3}a - 2r)^2 = 4r^2(a+p) - 2ar(\sqrt{3}a-2r)$, which yields $p = \frac{4ar^2+a(\sqrt{3}a-2r)^2-2ar(\sqrt{3}a-2r)}{3a^2-4\sqrt{3}ar} = \frac{3(a^2-2\sqrt{3}ar+4r^2)}{3a-4\sqrt{3}r} = \frac{(a-\sqrt{3}r)^2+r^2}{a-4r/\sqrt{3}}$. A calculator then gives us $p \approx 9.0187$ for $a = 10$ and $r = 1$.

Problem 2.25. The square $ABCD$ with side length $a = 8$ contains three pairwise tangent circles, with radii r, t and t, as shown in the figure. Determine $2r$ in terms of a.

Answer. $2r = \frac{9a}{16}$. If $a = 8$, then $2r = 4.5$. This problem was proposed by Watanabe Miya.

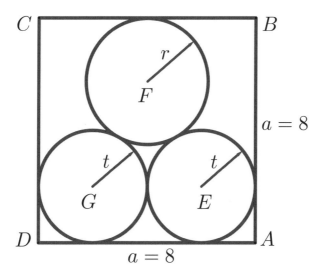

Solution by H. Fukagawa. It is easy to see that $t = \frac{a}{4}$ and $a = r + \sqrt{(r+t)^2 - t^2} + t$ hold, and from this, we obtain $2r = \frac{9a}{16}$.

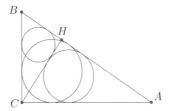

Problem 2.26. We are given a right triangle ABC with its altitude CH. The inradii of triangles ABC, ACH and BCH, are r, s and t, respectively. Show that $r + s + t = CH$ holds.

I could not find the name of the proposer of this problem.

Solution by H. Fukagawa. It is well known that $r = \frac{BC+AC-AB}{2}$, $s = \frac{AH+CH-AC}{2}$, $t = \frac{CH+BH-BC}{2}$ holds. From this, we immediately obtain $r + s + t = CH$. This problem is quite elementary.

Problem 2.27. We are given a circle of radius r tangent to both the side AD and the diagonal AC of a square $ABCD$. DE is also tangent to the circle. Determine $DE = p$ in terms of a and r.

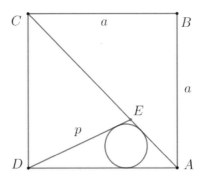

Answer. $p = \frac{a((k-r)^2+r^2)}{k^2-2r^2}$ for $k = a - \sqrt{2}r$. If $a = 169$ and $2r = 9.9$ then $p \approx 159.28455$. This problem was proposed by Nakkajima Motokae.

Solution by H. Fukagawa. The area of triangle AED can be calculated by

$$A(\triangle AED) = \frac{1}{2} aq \sin 45° = \frac{\sqrt{2}}{4} aq = \frac{1}{2}r(p+q+a)$$

for $AE = q$. We therefore have $\sqrt{2}aq = 2r(p+q+a)$ or $q = \frac{2r(a+p)}{\sqrt{2}a-2r} = \frac{2r(a+p)}{\sqrt{2}k}$ for $k = a - \sqrt{2}r$. Since

$$p^2 = a^2 + q^2 - 2aq \cos 45° = a^2 + q^2 - \sqrt{2}aq,$$

we have $p^2 - a^2 = q(q - \sqrt{2}a) = \frac{2r(a+p)}{\sqrt{2}k}\left(\frac{2r(a+p)}{\sqrt{2}k} - \sqrt{2}a\right)$. Dividing by $p + a$, we obtain $p - a = \frac{\sqrt{2}r}{k}\left(\frac{\sqrt{2}r(a+p)}{k} - \sqrt{2}a\right)$, or $(p - a)k^2 = 2r(r(a+p) - ak)$. Separating p then gives us

$$p = \frac{2r^2a + ak^2 - 2ark}{k^2 - 2r^2} = \frac{a((k-r)^2 + r^2)}{k^2 - 2r^2}.$$

If $a = 169$, $r = \frac{9.9}{2} = 4.95$, then $k = 169 - \sqrt{2} \times 4.95 \approx 162$ and $p \approx \frac{169\{(162-4.95)\}^2+4.95^2}{162^2-2\times4.95^2} \approx 159.285$. The result on the tablet is correct.

Problem 2.28. Two lines BE and FG are drawn as shown in a given square $ABCD$, such that they are tangent to three circles with radii R, r and t, respectively. Determine BE, FG, $2R$, $2r$ and $2t$ in terms of a.

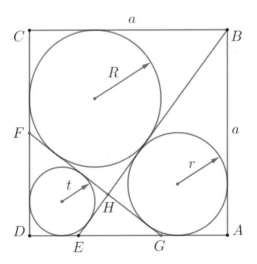

Answer. $BE = \frac{5}{4}a$, $FG = \frac{5}{6}a$, $2R = \frac{2}{3}a$, $2r = \frac{a}{2}$ and $2t = \frac{a}{3}$. On the tablet, the result $a + BE + FG + 2R + 2r + 2r + 2t = \frac{55}{12}a$ was written.

This problem was proposed by Nakata Sadakatsu.

Solution by Tameyuki Yoshida (1819–1892), translated by the author. This problem was quoted from the famous book, "Seiyou Sanpo" which was published in 1781.

A samurai and mathematician living in Nagoya, Tameyuki Yoshida included this original solution to his problem.

$FH = p = FD = FC$ implies $p = \frac{a}{2}$. The similar triangles $\triangle BCF \approx \triangle FDE$ give us $BC{:}CF = FD{:}DE$, and therefore with $HE = DE = q = \frac{p}{2} = \frac{a}{4}$ and $BH = a$, we have $BE = \frac{5}{4}a$ Finally, from $\frac{p}{a} = \frac{R}{a-R}$ in triangle FHB, we obtain $R = \frac{a}{3}$.

Problem 2.29. We are given a right triangle ABC. DE is tangent to the incircle of the triangle with $DE \| BC$ and $AD + DE + EA = k$. Determine BC in terms of k.

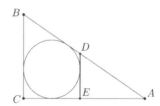

Answer. If $k = 6$, then $BC = 4$. This problem was proposed by Watanabe Makoto.

The author has not yet been able to solve this problem.

Problem 2.30. BH is perpendicular to the side AC of the triangle ABC. The incircle of ABC has the radius r. If $h = BH = 16$, $q = AH = 30$ and $r = 7$, determine $b = AC$ and $a = BC$.

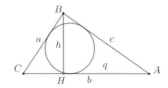

Answer. $b = AC = 42$ and $a = BC = 20$.

This problem was proposed by Watanabe Touko.

The author has not yet been able to solve this problem.

3. Sangaku Problems of Tashiro Shrine (1841)

Would you like to use any problems of this sangaku for your competition?

Five problems were drawn on this tablet.

Problem 3.1. This problem was proposed by Tsuchiya Fusakich, eleven years old, Tsuchiya Nobuyoshi school.

Problem 3.2. This problem was proposed by Iguchi Momoichiro, thirteen years old.

Problem 3.3. This problem was proposed by Hibino Heinojyou, eleven years old.

Problems 3.4 and 3.5 are quite difficult ones concerned with ellipses, and not well suited for most students. We will not consider these problems here.

This sangaku was hung in the year (1841) by the excellent mathematician Tani Yusai. Its size is 99×64 cm.

Problem 3.1. We are given a square with sides of length a. Determine the diameter x of a circle with the same area as the square, in terms of a.

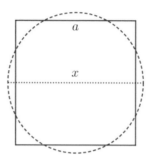

Answer. $x = 1.128379 \times a$. The sangaku issues a challenge to visitors: "Can you solve this? Determine the diameter x of the circle in terms of a."

Solution by H. Fukagawa. The area of the circle is $a^2 = \frac{\pi}{4}x^2$.

From this, we obtain the result $x = \sqrt{\frac{4}{\pi}} \cdot a \approx 1.128 \times a$.

Problem 3.2. In a square with sides of length a, we draw four circles with radius $r = \frac{a}{4}$ (and diameter $t = 2r$), touching externally. S is the area left in the interior of the square after removing the four circles. Determine the value of t in terms of S.

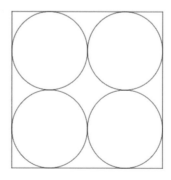

Answer. $t = \frac{1}{2}\sqrt{\frac{S}{\left(1-\frac{\pi}{4}\right)}}$.

Solution by H. Fukagawa. It is clear that $a = 2t = 4r$. We have

$$S = 4t^2 - 4 \times \pi r^2 = 4\left(t^2 - \pi r^2\right) = 4\left(t^2 - \frac{\pi}{4}t^2\right) = 4\left(1 - \frac{\pi}{4}\right)t^2,$$

which implies $t = \frac{1}{2}\sqrt{\frac{S}{\left(1-\frac{\pi}{4}\right)}}$.

Problem 3.3. Two lines TP and TQ pass through the vertices of four squares as shown in the figure, where three equally large squares with sides of length a have two sides each in common with a big square, and a small square with sides of length b has three common vertices with the three equally large squares. Determine b in terms of a.

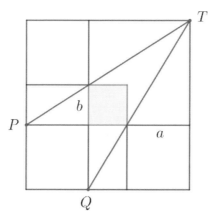

Answer. $b = \frac{\sqrt{5}-1}{2}a$.

Solution by H. Fukagawa. From the figure, the similar triangles on the right give us $\frac{2a+b}{a+b} = \frac{a+b}{a}$, which implies $2a^2 + ab = a^2 + 2ab + b^2$ and therefore $b = \frac{\sqrt{5}-1}{2}a$, since we have $b > 0$. The desired result therefore gives us $\frac{a}{b} = \frac{\sqrt{5}-1}{2}$, i.e. the Golden Section.

Bibliography

Fukagawa, H., and Pedoe, D. (1989). *Japanese Temple Geometry Problems: San Gaku*, Winnipeg, Canada: Charles Babbage Research Ctr.

Fukagawa, H., and Rothman, T. with foreword by Dyson, F. (2008). *Sacred Mathematics: Japanese Temple Geometry*, Princeton University Press, USA.

Chapter 1.5

Can We Pose Problems That are Attractive, Yet Accessible to Many?

Edmundas Mazėtis and Romualdas Kašuba

Vilnius University, Lithuania

Some problems are so simple, and their statements so natural, that one might have a rather strong feeling that they have always existed — similar to the sky or the Earth. In such a situation, it is not so easy to believe that these problems were actually created by some concrete person. One might then have a rather strong impression that the creation of such natural problems also has some of the remarkable flavor of true art.

One such problem is the following, taken from a math event for students of Grade 5, proposed in Belarus. This problem is such that all the essential components of its solution can be summarized in a single sentence.

Problem 1.1. The teacher asked her five students named Andrew, Basil, Chris, Daniel, and Elin to each multiply two digits from the multiplication table in the order mentioned. The students were happy to follow the wish of their teacher and did their calculations correctly.

If each of them got a result that was exactly one and a half times as big as the answer of the one before, what was the result that Daniel got?

The only thing we need to do here, is to find five results in the multiplication table, none of which is too big (note that only single digits are being multiplied), and the first of which is divisible by 2 sufficiently often. The numbers are therefore from 16 to 81, and Daniel got 54.

The second problem we present here is excellently suited to any student with enough patience to do simple arithmetic step by step, gradually improving his results.

Problem 1.2. A grandmother with five granddaughters has some sweets to distribute between them. Loving puzzles almost as much as her granddaughters, she arranges the distribution of sweets in such a way that any two girls together get a different total number of sweets. Simultaneously, any three of them together get more sweets than the remaining two. What is the least possible number of sweets that Grandma needs for her trick?

As is often the case with problems of this type, we can start out with an example of such a distribution. We might then be able to construct something by improving this initial example.

A first example of this type could be 7, 10, 13, 14 and 15, for instance. In this case, the 10 possible sums of pairs are

$$17, 20, 21, 22, 23, 24, 25, 27, 28 \text{ and } 29,$$

and these sums are indeed all different. Furthermore, the sum of the three smallest numbers, namely 30, is larger than the sum of the two biggest numbers, which is 29. In this case, the sum of all five numbers is 59.

Now, looking for smaller numbers and their different pairwise sums is rather similar to a process of going down steps, where you might came across smaller numbers fulfilling the conditions along the way. An example could be

$$6, 10, 12, 13 \text{ and } 14,$$

five numbers for which all pairwise sums are again different, with the sum of the smallest 3 greater than the sum of the remaining two. For these numbers, the sum is equal to 55, which is already less than 59. Continuing your efforts you might then find another 5 suitable numbers, like

$$6, 9, 11, 12 \text{ and } 13,$$

for which the sum is only 51. After this, any further attempt at improving the value will not seem to work. It is therefore then time to prove that 51 is indeed the lowest possible value, and this proof is left to the interested reader.

Let us now consider another problem originating in Belarus.

Problem 1.3. The number N is said to be *more-than-smart* if it contains all 10 digits, is divisible by 9 and has the following property: If a digit a is bigger than a digit b, then the digit a appears more often in the decimal representation of N than the digit b. Determine the smallest possible more-than-smart number.

Here, we must deal with very large numbers, and a possible first step in the search for a solution turns our attention to the number

10122233334444455555566666667777777788888888899999999999.

The sum of the digits in this number is equal to

$$1 \cdot 2 + 2 \cdot 3 + 3 \cdot 4 + 4 \cdot 5 + 5 \cdot 6 + 7 \cdot 8 + 8 \cdot 9 + 9 \cdot 10 = 330.$$

330 is clearly not divisible by 9, and we must therefore increase the number in the smallest possible way in order to gain divisibility by 9 without losing the monotonicity of its digits. Once again, this last step is left to the interested reader.

The following especially nice problem is taken from a Swedish source of creative mathematics.

Problem 1.4. A young man enters a 7-Eleven shop and wants to buy 4 items. The reluctant shop assistant demonstratively multiplies the prices of these items and solemnly declares that the customer has to pay 711 cents, i.e. 7 dollars and 11 cents. The young man is slightly irritated by this calculation and asks since when prices of individual items in a purchase are multiplied. The unshakable shop assistant then adds the prices and announces that the customer has to pay exactly the same amount as stated before, namely 7 dollars and 11 cents. What are the prices of the individual items?

The full solution is again left to the interested reader, but note that there are not many ways to obtain the number 711 as the product of four integers!

An important mathematical tool is the extraction of information from various kinds of data presented in tables. On the other hand, we are also often confronted with problems in which we are asked whether it is possible to construct a table fulfilling some prescribed, and sometimes rather sophisticated, properties. With this in mind, let us continue with the following problem related to the filling of tables and data preparation. This is yet another problem taken from a Belarusian math contest.

Problem 1.5. We consider a 3×4 table and set out to fill all 12 of its cells with distinct positive integers in such a way that the sum of the numbers in each row is equal and the sum of the numbers in each column is equal. Determine the smallest possible sum of the 12 integers.

You cannot say that this is a very difficult task. At the same time, we dare to state that this problem is not one that might easily be solved in 60 seconds.

It is not at all difficult to find 12 distinct numbers we can place in a table fulfilling the required conditions. Considering distinct numbers with equal sums in the rows, we immediately see that the sum of all numbers must be divisible by 3. Similarly, considering the four columns, we see that the sum of all numbers must be divisible by 4. We therefore see that the sum of all numbers must be divisible by 12.

After a few seconds' thought, this might lead us to have a look at the sum of the first 12 natural numbers, which is 78 and by no means divisible by 12. This indicates that the smallest possible sum of a distinct dozen fulfilling the required conditions might be 84. It remains only to look for an example showing that the sum 84 can indeed be achieved. An example of such a table is presented below.

1	13	10	4
11	2	3	12
9	6	8	5

A close look shows us that our table contains all natural numbers from 1 through 14 with the exception of the middle number 7. Moreover, each row contains two pairs of integers with the sums 14, i.e. the pairs (1, 13) and (4, 10) in the first row, (11, 3) and (2, 12) in the second, and the two remaining pairs (9, 5) and (6, 8) in the third.

Now, let us take a look at another accessible and also rather interesting problem concerning the sums in the even smaller 3×3 table.

Problem 1.6. We would like to write (not necessarily distinct) positive integers in the cells of a 3×3 table in such a way that the 3 sums of the numbers in the rows of the table and the 3 sums in its columns are six different primes. Determine the smallest possible sum of the numbers in the table.

A preliminary observation we might make concerning the parities of the numbers in the table immediately eliminates 2 as a possible sum. The 6 smallest remaining primes are then 3, 5, 7, 11, 13, and 17, but we can readily show that these sums are not possible. Adding them and then dividing by 2 gives us the result 28, which can be shown to be impossible. If we then replace 17 by 19, it will turn out that it is possible to create a table fulfilling the required properties in this case. Again, the details are left to the interested reader.

Sometimes, it can happen that the creation of a table with certain specific nice properties is more difficult than expected. This might have several reasons. For one, the required properties of the table may be very tersely stated. On the other hand, the solutions given in books are often confusingly short. Such problems may seem rather easy at first glance, but the results of the contestants attempting such a problem at a competition will prove otherwise. Below we would like to present an example of this kind. The problem was proposed for the younger grades in the Lithuanian Mathematical Olympiad, where only very few solutions were found by the participants.

Problem 1.7. A number is written in each cell of a 7×7 table in such a way that the product of the 9 numbers in any 3×3 square is equal to the product of the 16 numbers in any 4×4 square (and equal to a given number S). Is it possible that there exists a value of S such that the product of all 49 numbers in the initial table can be equal to 2017?

As mentioned, this problem was thought to be quite accessible when it was first proposed. How will the interested reader fare?

Sometimes, it is remarkably good for the sake of extending our students' boundaries of thought to pose a slightly provocative — or just challenging — problem with a seemingly clear answer, and require a clear explanation as to why there is no other answer.

In other words, it might happen that in dealing with the problem, you feel immediately sure that you know the answer, but at the same time you might not be able to guarantee it. One of the best examples illustrating such a situation is the following task of Russian origin. We have allowed ourselves the freedom to join the word Graz to this excellent and classic exercise, in honor of our 2018 meeting.

Problem 1.8. A natural number N is said to be *smart* if its digits from left to right are strictly increasing — like the number 12, which is the smallest of the smart numbers. We define the *Graz estimation* of a smart integer to be the sum of the digits of the integer $9N$. What are the possible values of the Graz estimations of all smart integers?

From first sight, any even slightly skilled problem solver has the unshakable realization that the Graz estimation of any smart integer — as the sum of the digits of a multiple of 9 — must be divisible by 9.

For the example of the smallest smart integer 12, we have $9 \cdot 12 = 108$, and its Graz estimation is therefore $1 + 0 + 8 = 9$. You can try any other example, and you might then predict that you will get 9 again. Trying again and again, we get nothing but 9. But why is this so?

And so, we ask you, the reader to decide. Is 9 indeed the only possible result and if so, what are the reasons for that?

The International Mathematical Olympiad 2016 took place in Hong Kong. In order to honor this, a problem series with a Chinese flavor was prepared for the Lithuanian contests for younger grades (Grades 5–8) which were organized simultaneously with the Lithuanian team contest. Here are two problems from this problem set.

Problem 1.9. The spirit of Father Ming is still alive, it seems. To honor him, four grandchildren of the famous man wish to split the square drawn below into four congruent connected parts. Each of the parts should contain exactly four digits that could then be combined to form the number 2016. Will they be successful in achieving their goal and honoring the beloved Father Ming? We think that with your gracious help they will be able to deal successfully with this task.

	2				
	2	6	2		
	1	0	1	0	
	0	1	0	1	
		6		6	
2				6	

One possible answer is as shown, with the four parts marked with A, B, C and D:

B	2B	A	A	A	A
B	2A	6A	2C	C	A
B	1A	0A	1C	0C	C
B	0B	1B	0D	1D	C
D	B	6B	D	6D	C
2D	D	D	D	6C	C

Another problem from this set is related to the splitting of a regular hexagon into triangles.

Problem 1.10. The famous Chi Gen Geo, who historians claim was a real person and not mythical, felt the strongest attraction to solving all kinds of geometrical puzzles from his early childhood. Many of his puzzles were considered unsolvable by his compatriots. Because of this, he entered adulthood surrounded by unceasing admiration and fame. His most unforgettable statement held that a properly made drawing of any geometrical problem is the solid bridge joining the Continent of Necessary Efforts and the Land of Achieved Successes. It is said that, when Chi Gen Geo was still younger than 10, he was able to solve the problem below in 10 minutes. Moreover, it is said that he used no paper in his solution.

We are given a regular hexagon *ABCDEF* with area 1. Let us consider all possible triangles with vertices among the vertices of the hexagon. What is the total of the areas of all these triangles?

In order to solve this problem, it is rather useful to get some idea about what kind of triangles we might get by choosing three vertices of the hexagon. Simple analysis leads us to the fact that all such triangles are of exactly three sizes.

Triangles of the first type are obtained by connecting any three adjacent vertices of the hexagon.

As an example the triangle *ABC* might be taken into consideration. There are also five other such triangles, namely *BCD*, *CDE*, *DEF*, *EFA* and *FAB*, and so there are six triangles of this first type.

Triangles of the second type are obtained by connecting alternate vertices of the hexagon. There are only two such triangles, namely *ACE* and *BDF*.

We can immediately observe that each triangle of the second type can be obtained by cutting three triangles of the first type from the hexagon. This implies that the common area of three triangles of the first type together with one triangle of the second type is equal to the area of the hexagon, which is 1, and the six triangles of the first type together with the two triangles of the second type therefore cover the original hexagon exactly twice.

There are also triangles of a third type which we obtain by connecting any two adjacent vertices with a third, non-adjacent one. No other triangles exist, other than these three types. It therefore only remains to determine what part of the hexagon is covered by the triangles of this third type.

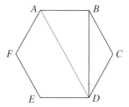

In order to calculate the area of a triangle of the third type, we make the following observations.

(1) The regular hexagon with sides of length *a* can be divided into six equilateral triangles of the same side length by drawing the three diagonals of the hexagon.
(2) Two such adjoining equilateral triangles form a rhombus, whose area is one third of the area of the hexagon.
(3) Any such a rhombus can be cut into two triangles of the first type by drawing the longer diagonal. It follows that the area of such a triangle is one-sixth of the area of the hexagon.
(4) We now note that *ABDE* is a rectangle that results by removing two such triangles from the hexagon, and therefore covers two-thirds of the area of the hexagon. Each triangle of the third type covers half of that rectangle, or one-third of the area of the hexagon.

It remains to note that there exist exactly 12 triangles of the third type, two with each of the six sides of the hexagon as their shortest sides. All 12 such triangles together therefore cover the hexagon exactly 4 times, and adding this to the two coverings we have already determined by triangles of the first two types, we see that the total area of all triangles is equal to 6 times the area of the hexagon.

In the following section, we would like to present some additional geometrical problems. In each of them, it is essential to decide where to add a line segment to the drawing and how the sizes of inscribed angles are related to another.

Many things can be said about the importance of illustrations in geometry. We have always tried to make the importance of illustrations comprehendible for our students. In our attempts to clarify this, we found the following comparison useful: The importance of a drawing in a geometric problem is not much less than that of a wedding gown during a wedding ceremony.

Problem 1.11. Let $ABCD$ be a square. A line l is drawn through the point C perpendicular to AC, and points E and F are marked on that line in such a way that $BDFE$ is a rhombus. Determine the angle CBF

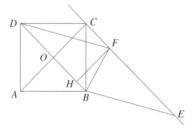

Let O be the point of intersection of the diagonals of the square. Draw the line FH perpendicular to BD with $H \in BD$. We note that BD is perpendicular to OC and $FH = OC = \frac{1}{2}BD = \frac{1}{2}DF$. Since HDF is a right triangle, we therefore have $\angle HDF = 30°$. Since the triangle BDF is isosceles, we therefore have $\angle DBF = \frac{1}{2} \cdot (180° - 30°) = 75°$, and since $\angle DBC = 45°$, we obtain $\angle CBF = \angle DBF - \angle DBC = 30°$.

Problem 1.12. The circle S_1 with the center O_1 and the circle S_2 with the center O_2 intersect at points A and B. A point C is marked on the arc of S_1 inside the circle S_2. The lines AC and BC intersect the circle S_2 a second time at points E and D, respectively. Determine the angle between the lines DE and O_1C.

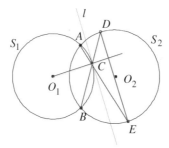

To solve this, it is sufficient to add the tangent line l through C to the first circle. The angle between l and AE is the same as the angle ABC. Since $\angle ABC = \angle AED$, this means that $l \parallel DE$ holds, and therefore $O_1C \perp DE$.

The next problem was also proposed at the Lithuanian team contest for youngsters. This problem is really democratic in the sense that any skilled teacher and any smart student will require approximately the same amount of time to find a solution.

Problem 1.13. Alice, accompanied by her bodyguard Cheshire Cat, wishes to use each of the digits 2, 3, 4, 5, 6, 7, 8 and 9 once to fill in the boxes in the equation below to make it correct. Of the three fractions being added, what is the value of the largest one?

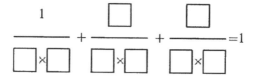

This problem is of Chinese origin, and is remarkable in that it looks so simple at first glance and yet is not at all easy to solve. Taking a closer look at the numbers 2, 3, 4, 5, 6, 7, 8 and 9, and thinking about where some of them could be placed, we note that the digits 5 and 7 are all alone in the list, as they have no multiples accompanying them there. This means that they cannot be located in the denominators. After discovering this, the rest or the proof becomes rather straightforward. We can simply multiply out and get rid of the denominators. The digits 5 and 7 must be located in the numerators and the rest is then a technical matter, left to the interested reader. Noting that the LCM of the remaining numbers 2, 3, 4, 6, 8, and 9 is 72, we quickly arrive at the solution.

The next few problems involve some nice number properties.

Problem 1.14. Four intersecting circles enclose a total of 10 regions in the plane, as shown in the figure.

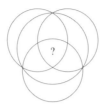

We would like to place the numbers 1 through 10 in their interiors, with one number in each of the regions. This is to be done in such a way that adding the numbers inside every circle gives the same sum for all four circles.

What number should then be placed in the region containing the question mark?

(A) 1   (B) 2   (C) 4   (D) 6   (E) 7

Problem 1.15. The rectangle *ABCD* is divided into squares as shown. The length of the side *AB* is 32.

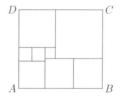

What is the length of the side *AD*?

(A) 26   (B) 27   (C) 28   (D) 29   (E)   30

Problem 1.16. You may choose any number consisting of five different digits. The digit 0 may not be used. In a next step, you may choose two adjacent digits from this number and switch their positions. You may perform this step five times in total. Finally, you are to compute the difference between the chosen number and the final number obtained after switching. What is the largest possible difference that you can possibly obtain?

For example one may start with the number 45128. After switching, you may obtain, in this order, the numbers 45182, 45812, 48512, 84512 and 85412. Then, the difference between the initial number and the final number is $85412 - 45182 = 40230$.

Problem 1.17. Matches of the lengths 8 cm and 9 cm each are given, with their total length equal to 1800 cm Construct a regular octagon using all of them without breaking any of the single matches. This problem was first suggested in Saint Petersburg.

Sangaku geometry (note also Chapter 1.4 on sangaku problems in this volume) is never too easy but always exciting. Maybe this is because there does not ever seem to be too much empty space left in any sangaku drawing, or because they give the impression that some geometrical adventure might happen anywhere. Here are a few such problems.

Problem 1.18. Let $ABCD$ be an isosceles trapezoid, with $AB \parallel CD$ and $CD > AB$. The two circles ω and ω_1 have equal radii a with centers E and E_1, are tangent at F, and each touch the segment CD. The circle ω is also tangent to the segment AD, and the circle ω_1 to the segment BC. The two circles γ and γ_1 have equal radii c with centers H and H_1, are tangent at I, and each touch the segment AB. The circle γ is tangent to segment AD, and the circle γ_1 to the segment BC. The circle θ with center G is tangent to

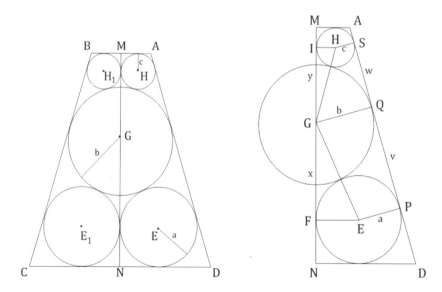

circles ω, ω_1, γ, γ_1 and segments AD and BC (as can be seen in the figure of this problem). Determine the radius b of θ.

Let us enjoy the solution. Let M and N be the midpoints of sides AB and CD. Because of the symmetry across the line MN, the sides of the trapezoid $AMND$ touches circles γ and ω. Also, G lies on MN (as in the figure on the right). We draw $EF \perp MN$, $HI \perp MN$, $EP \perp AD$, $GQ \perp AD$, $HS \perp AD$, GE and GH. Since $EF = a$, $HI = c$ and $GQ = b$, we have $EG = a + b$, and $GH = b + c$. Let $FG = x$, $GI = y$, $QP = v$, $QS = w$. Let lines AD and MN intersect at point U. Since $UI = US$ and $UF = UP$, we have $IF = IP$ and $v + w = x + y$. From the triangles EFG and GHI we have $x = \sqrt{(a+b)^2 - a^2} = \sqrt{2ab + b^2}$, $y = \sqrt{(b+c)^2 - c^2} = \sqrt{2bc + b^2}$. Since $v = 2\sqrt{ab}$, $w = 2\sqrt{bc}$, from equation $v + w = x + y$ we obtain the equation $2(\sqrt{a} + \sqrt{c}) - \sqrt{b + 2c} = \sqrt{b + 2a}$. After some calculation, we get $3c + a + 4\sqrt{ac} = 2(\sqrt{a} + \sqrt{c})\sqrt{b + 2c}$. Since $3c + a + 4\sqrt{ac} = (3c + 3\sqrt{ac}) + (a + \sqrt{ac}) = (\sqrt{a} + \sqrt{c})(\sqrt{a} + 3\sqrt{c})$, we have $\sqrt{a} + 3\sqrt{c} = 2\sqrt{b + 2c}$, and $b = \frac{a + 6\sqrt{ac} + c}{4}$.

Problem 1.19. In a semicircle with diameter AB, the two equal chords $AP = BQ = d$ are each tangent to two small equal circles as shown in the figure. Determine the length of AB and the radius of the small circles.

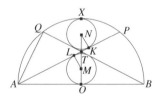

Solution. Note that this problem is an extension of Problem 2.10 from Chapter 1.4. Let lines AP and BQ intersect at T, and let M and N denote the centers of the small circles, and O the midpoint of AB. Let the circle with center N be tangent to the semicircle with diameter AB at point X, the line AP be tangent to the circle with center N at point K, and the circle with center M at point L. Since $ML = NK$ and $ML \parallel NK$, then $\triangle NKT = \triangle MLT$. Therefore $MT = NT$ and $OT = TX = \frac{AB}{4}$. Since $\triangle ABQ \sim \triangle TBO$, we have $\frac{AB}{QB} = \frac{TB}{OB}$. Therefore

$$AB = \frac{TB \cdot QB}{OB} = \frac{d}{0.5AB}\sqrt{OT^2 + OB^2} = \frac{d}{0.5AB}\sqrt{\frac{AB^2}{16} + \frac{AB^2}{4}} = \frac{d}{0.5AB}.$$

$\frac{\sqrt{5}}{4}AB = \frac{d\sqrt{5}}{2}$. Since $NK \perp AP$, $NT \perp AB$, we have $\angle KNT = \angle ABQ$. Therefore $\triangle KNT \sim \triangle QBA$, and $\frac{NK}{TN} = \frac{QB}{AB} = \frac{2}{\sqrt{5}}$. Thus $\frac{NK}{\frac{AB}{4} - NK} = \frac{2}{\sqrt{5}}$.

and the radius of the small circles is equal to $NK = \frac{d(5 - 2\sqrt{5})}{4}$.

Problem 1.20. Two equal circles each touch two sides of a square and they also touch each other at the center of the square. Each of two smaller circles touches two sides of the square as well as a common tangent of the two larger circles as shown in the figure. Determine the radius of the smaller circles, if the length of the side of the square is a.

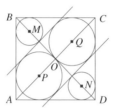

Solution. Let $ABCD$ denote the vertices of the square and O its center. Furthermore, let P and Q denote the centers of the larger circles, and M and N the centers of the smaller circles. Let R and r denote the radii of the larger and smaller circles, respectively. Since $AP = R\sqrt{2}$ and $AP = \frac{1}{2}(AC - 2R)$, we have $R\sqrt{2} = \frac{1}{2}(a\sqrt{2} - 2R)$, which yields $R = \frac{\sqrt{2}a}{2 + \sqrt{2}}$. Since $DM = r\sqrt{2}$ and $DM = \frac{1}{2}(BD - 2R - 2r)$, we obtain $r = \frac{a\sqrt{2} - 2R}{2\sqrt{2} + 2} = (3 - 2\sqrt{2})a$.

The following four problems are all quite natural problems concerning semicircles. As will be the case for most of the problems presented in the final part of this chapter, these are presented without solutions, in the hope that interested readers will be motivated to find their own solutions.

Problem 1.21. The segment AB is a diameter of a semicircle. Points S and T lie on the semicircle with $\angle SAT = \alpha$. The point M is the midpoint of segment ST. Let P be the point on AB such that $SP \perp AB$. Determine the angle SPM.

Instead of the complete solution, it should be enough to mention here that if O is the midpoint of AB, we have $OM \perp ST$. This implies that the quadrilateral $POMS$ is cyclic and therefore $\angle SPM = \angle SOM = \frac{1}{2}\angle SOT = \angle SAT = \alpha$.

Problem 1.22. Points C and D lie on a semicircle with diameter AB such that A, D, C, B lie on the semicircle in that order. The tangents to the semicircle at points C and D meet at E, the segments AC and BD meet at F, and the segments EF and AB meet at M. Prove that the points E, C, M, D are concyclic.

Problem 1.23. Let the point M be the midpoint of a segment AB. We are given two semicircles: ω with diameter AB and γ with diameter MB, with semicircle γ inside semicircle ω. The points X and Y lie on γ such that length of the arc BX is 1.5 times the length of the arc BY. The line MY intersects the line BX and semicircle ω at points D and C respectively. Prove that $CY = YD$ holds.

Problem 1.24. Segments AD and BE are altitudes of an acute triangle ABC. Two semicircles ω_1 and ω_2 have the diameters BC and AC, respectively, and lie outside the triangle. The line AD intersects the semicircle ω_1 at a point P, and the line BE intersects ω_2 at a point Q. Prove that $CP = CQ$ holds.

We start the solution with the observation that since PD is the altitude in the right triangle PBC, we have $PC^2 = CD \cdot CB$. Similarly, from the right triangle QAC we obtain $QC^2 = CE \cdot CA$. Since the right triangles BCE and ACD are similar, we have $\frac{CD}{CE} = \frac{CA}{CB}$, or $CD \cdot CB = CE \cdot CA$, and thus $PC^2 = QC^2$.

Next, we consider some problems, in which the consideration of inscribed quadrilaterals will prove to be a quite useful tool for angle chasing.

Problem 1.25. In the quadrilateral $ABCD$ with $\angle DAB = 110°$, $\angle BCD = 35°$, and $\angle CDA = 105°$, the diagonal AC bisects the angle DAB. Determine the angle ABD.

Let the point I be the incenter of triangle ABD. The point I must then lie on the segment AC. Since we have $\angle BID = 90° + \frac{1}{2}\angle BAD = 90° + 55° = 145°$ and $\angle BID + \angle BCD = 180°$, the quadrilateral $BIDC$ is cyclic. Therefore, we have $\angle IBD = \angle ICD = 180° - \angle ADC - \angle DAC = 180° - 105° - 55° = 20°$, and thus $\angle ABD = 2\angle IBD = 40°$.

Problem 1.26. In the acute triangle ABC, BD and CE are altitudes. The line ED intersects the circumcircle of ABC at P and Q. Show that $AP = AQ$.

Solution. Since $\angle BDC = \angle BEC = 90°$, we see that $BCDE$ is cyclic. Hence, $\angle AEQ = \angle ACB$. Now, we note that $\angle APQ = \angle AEQ - \angle PAB = \angle ACB - \angle PQB$ holds. On the other hand, we also have $\angle AQP = \angle AQB - \angle PQB$. Since $\angle ACB = \angle AQB$ holds, we conclude $\angle APQ = \angle AQP$, and hence $AP = AQ$.

Next, we offer a nice mixture of geometric problems of various types that have caught our fancy recently. For the first few, the application of transformations is a natural tool that can prove to be quite useful. All of these problems have some kind of an interesting twist, and are well worth spending some time on. Most are offered without a solution, in the hope that interested readers will enjoy the discoveries to be made on their own.

Problem 1.27. We are given a rectangle $ABCD$ with $AB = 9$ and $BC = 8$. Points E and F lie inside the rectangle, such that $EF \parallel AB, BE \parallel DF, BE = 4$, and $DF = 6$ hold, with E is closer to BC than F. Determine the length of EF.

Solution. Let M and N be the points on segments DC and AB, respectively, such that $CM = BN = EF$. We consider the translation with vector $\overrightarrow{FE}$. The images of points C and B under this translation are the points M and N, and the image of the point F is E. Since $BE \parallel DF$, the points D, F and N are collinear and we have $DN = DF + EB = 10$. If $EF = x$, we have $AN = 9 - x$. The Pythagorean theorem in the right triangle ADN gives us $(9 - x)^2 + 8^2 = 10^2$. Therefore, we have either $x = 3$ or $x = 15$. Since $x < 9$ (the segment EF lies inside the rectangle $ABCD$), we have $EF = 3$.

Problem 1.28. Given a right triangle $ABC, \angle C = 90°$, let $ABDE$ be a square external to ABC. Prove that the bisector of the angle in C bisects the area of the square $ABDE$.

Solution. Let O be the center of the square $ABDE$. Since $\angle BCA = \angle BOA = 90°$, the points A, C, B, O are concyclic. Since $AO = OB$, the point O is the midpoint of arc AB. It therefore follows that the line CO bisects the angle C. Since the square is centrally symmetric, the lines that bisect the area of square pass through its center. It follows that the line CO bisects the area of the square $ABDE$, as claimed.

Problem 1.29. Let $ABCD$ be a parallelogram with $AB > AD$. Points P and Q are given on the sides AB and AD, respectively, such that $AP = AQ = x$. Prove that as x varies, the circumcircle of the triangle PQC always passes through another fixed point (other than C).

Problem 1.30. Let I be the incenter of a triangle ABC, and let A_1, B_1, and C_1 be the reflections of I across the lines BC, CA, and AB, respectively. The circumcircle of the triangle $A_1 B_1 C_1$ passes through the point A. Determine $\angle A$.

Problem 1.31. Two circles intersect in points M and N. The vertices A and C of the rectangle $ABCD$ lie on the first circle, and the vertices B and D lie on the second circle. Prove that the point of intersection of the diagonals AC and BD lies on the line MN.

Solution. The line MN is the radical axis of the circles. Since $ABCD$ is a rectangle, it is cyclic. Applying the radical lemma, the segments AC, BD and the radical axis MN intersect in a common point.

Problem 1.32. The points E and F lie on the extensions of the sides AD and CD of the square $ABCD$ beyond D with $AB = 1$ and $DE = DF = \frac{1}{2}(\sqrt{5} - 1)$. Prove that all diagonals of the pentagon $ABCEF$ are parallel to the respective side of the pentagon.

Problem 1.33. $ABCD$ is a trapezoid with $AD \parallel BC$, $AD = 4$ and $BC = AB = 2$. Determine the angle ACD.

It is sufficient to mention that if O is the midpoint of the segment AD, then $AOCB$ is rhombus, since $AO = AB = BC$, and $AD \parallel BC$. It therefore follows that $OC = 2$ must hold. Since $OA = OD = OC$, CD is a right triangle with $\angle ACD = 90°$.

Problem 1.34. We are given a triangle ABC with $\angle A - \angle B = 90°$. The point E is the midpoint of segment AB. Let D be the foot of the altitude from C to AB. Prove that the radius of the circumcircle of the triangle ABC is equal to DE.

Problem 1.35. Let O and H be the circumcenter and the orthocenter of an acute angled triangle ABC, respectively. Let M be the midpoint of BC. Show that $AH = 2OM$ holds.

Solution. If we draw the diameter BD, the point O is the midpoint of BD. We have $OM = \frac{1}{2}CD$. Since $AD \perp AB$, $CH \perp AB$, we also have $AD \parallel CH$. Similarly, we also obtain $DC \parallel AH$ and the quadrilateral $ADCH$ is therefore a parallelogram. Therefore, we have $DC = AH$, which completes the proof.

Problem 1.36. Let $ABCD$ and $AKMN$ be parallelograms such that the point K lies on the segment BC, and D lies on the segment MN. Show that the areas of the parallelograms are equal.

Solution. We consider the triangle ADK. It shares the base AD and the corresponding altitude with parallelogram $ABCD$, and we therefore have $S_{ADK} = \frac{1}{2}S_{ABCD}$. Similarly, it also has the same base AK and altitude as parallelogram $AKMN$, and so we also have $S_{ADK} = \frac{1}{2}S_{AKMN}$. Therefore, $S_{ABCD} = S_{AKMN}$.

Problem 1.37. AD is an altitude of triangle ABC. The line through the point A parallel to BC intersects the circumcircle of the triangle ABC for the second time at E. Prove that the line DE passes through the centroid of triangle ABC.

Solution. Let points M and N be the midpoints of the segments BC and AE, respectively, and the lines AM and DE meet at point F. Since $MN \perp BC$, it follows that $DM = AN = \frac{1}{2}AE$ holds. Triangles AEF and MDF are similar, and we therefore have $\frac{AF}{FM} = \frac{AE}{MD} = 2$. It therefore follows that F is centroid of ABC, and we are done.

Problem 1.38. $ABCD$ is a trapezoid with $AD \parallel BC$, $AB = 10$, $CD = 24$, and $\angle A + \angle D = 90°$. Determine the distance between the midpoints of the sides AD and BC.

Solution. $AB < CD$ implies $\angle A > \angle D$. Let M and N be the midpoints of the sides AD and BC, respectively. Draw $NF \parallel AB$, $F \in AD$, $NE \parallel CD$, $E \in AD$. Since $\angle NFE = \angle A$, $\angle NEF = \angle D$, this implies $\angle FNE = 90°$ and $EF = \sqrt{NF^2 + NE^2} = \sqrt{AB^2 + CD^2} = 26$. Since $AF = ED = \frac{1}{2}BC$, M is the midpoint of EF. Therefore, we have $MN = ME = MF = 13$.

Problem 1.39. The point D is the midpoint of AB in a triangle ABC, with $\angle A = 30°$. Furthermore, we have $\angle CDB = 45°$. Determine $\angle B$.

Solution. Since $\angle A = 30°$, and $\angle CDB = 45°$, we have $\angle CDA = 135°$, and $\angle ACD = 15°$. We draw the circle with center D and radius AD, and let E denote the point in which it meets the segment AC. Since $\angle AEB = 90°$, we have $\angle ABE = 60^0$ and $EB = \frac{1}{2}AB = BD$. Therefore, the triangle BED is equilateral. We therefore have $\angle EDB = 60°$ and $\angle CDE = \angle EDB - \angle CDB = 60° - 45° = 15° = \angle ACD$. The triangle CED is therefore isosceles, with $CE = ED = EB$, and the right triangle CEB is also isosceles, from which we obtain $\angle ECB = \angle EBC = 45°$. Therefore $\angle B = \angle ABE + \angle EBC = 105°$.

Problem 1.40. Let the point H be the orthocenter of a triangle ABC with $AB = CH$. Determine the angle ACB.

Problem 1.41. The point S is the incenter of a triangle ABC, and the points A_1, B_1, C_1 are the reflections of S across the lines BC, AC, AB respectively. The circumcircle of the triangle $A_1 B_1 C_1$ passes through the point B. Determine the angle ABC.

Solution. Let r be the inradius of triangle ABC, and let $SK \perp BC$, and $K \in BC$. Then we have $SA_1 = 2SK = 2r$. Similarly, $SB_1 = SC_1 = 2r$. Therefore, the point S is the circumcenter of the triangle $A_1 B_1 C_1$. Since the circumcircle of the triangle $A_1 B_1 C_1$ passes through the point B, we have $SB = r$. In the right triangle SKB, we have $SB = 2r$, and $SK = r$, and thus $\angle SBK = 30°$. Since the line BS bisects the angle ABC, we have $\angle ABC = 60°$.

Finally, let us conclude with a few problems concerning some well-known heroes.

Problem 1.42. Billy Boy is especially fond of any 11-digit positive integer whose decimal expression contains only 0's and 1's, which ends in the digits 11, and which is divisible by 11. He respectfully refers to any such number as a number of *modest dignity*.

(a) Find an example of such a number of *modest dignity*.
(b) Determine the greatest number of *modest dignity*.
(c) Determine the smallest number of *modest dignity*.

Problem 1.43. Once upon a time, eight Cheshire brothers were participating in a chess competition, which had 7 rounds. The score was distributed according to the well-known rules of those days: 1 point for winning the game, ½ point for a draw and 0 points for the lost game.

(a) Show that after the first, the second, and the third round, at least two Cheshire brothers will have the same score.
(b) In the final ranking, no two Cheshire brothers have the same score. The eldest Cheshire brother always wins. Determine the smallest number of points the eldest brother could have gathered.

Problem 1.44. Alice is the most confident adviser of the Queen of Squares. The Queen used to discuss some challenging problems of value and importance with her. Once, the Queen told Alice:

> Take into account that E is the midpoint of the side CD of a square $ABCD$ and imagine yourself staying at some miraculous point M inside the square such that all three angles $\angle MAB$, $\angle MBC$ and $\angle BME$ — can you ever believe it? — are of the same magnitude. In case you are able to imagine such a situation, I, the Queen of Squares, order you to find the magnitude of those angles within the next three days — no more time will be given! — and announce it to the whole Kingdom — naturally with proof!

Bibliography

Andreescu, T., Rolinek, M., and Tkadlec, J. (2013). *107 Geometry Problems from the Awesome Math Year – Round Program*, XYZ Press, Dallas, USA.

Barabanov, E. A. *et al.* (Eds.) (2014). *Problems of the Belarus Math Olympiads*, School Years 2012/2013 and 2013/14, Minsk: "Konkurs" ass., p. 368.

Barabanov, E. A. *et al.* (Eds.) (2016). *Problems of the Minsk City Olympiad of Younger Grades* (2015–2012), Minsk: "Konkurs" ass., p. 302.

Barabanov, E. A. *et al.* (Eds.) (2017). *Problems of Minsk Regional City Olympiad* (2002–2011), Minsk: "Konkurs" ass., p. 256.

Chen, E. (2016). *Euclidean Geometry in Mathematical Olympiads*, MAA Press, Providence, USA.

Kashuba, R. (2012). *Kak reshat' zadachu, kogda nie znajesh kak* (*How to Solve the Problem When You Do Not Know How*), Moscow: Prosveshchenie Publishing House (2nd edition, 2014).

Kašuba, R. (2006). *What To Do When You Don't Know What To Do?* Rīga: Mācību grāmata.

Kašuba, R. (2007). *What To Do When You Don't Know What To Do? Part II*, Rīga: Mācību grāmata.

Kašuba, R. (2009). *Once Upon a Time I Saw the Puzzle Part III*, Rīga: University of Latvia.

Mazanik S. A. *et al.* (Eds.) (2005). *Zadachi minskoj gorodskoj olimpiady mladshykh shkolnikov* (*Problems of the Minsk City Math Olympiad for Younger Grades*), Minsk: "Konkurs" ass.

Rothman, T., and Fukagawa, H. (2008). *Sacred Mathematics — Japanese Temple Geometry*, Princeton: Princeton University Press.

Prasolov, V. V. (2006). *Problems in Plane Geometry*, Moscow: Nauka.

Chapter 1.6

A Functional Equation Arising from Compatibility of Means

Marcin E. Kuczma

University of Warsaw, Poland

1. Introduction

The following functional equation is the object of this presentation:

$$x + g(y + f(x)) = y + g(x + f(y)); \quad f, g : \mathbf{R} \to \mathbf{R}. \tag{1}$$

Before telling how it has emerged, let us try to get some feeling for the equation. It involves two unknown functions; a *solution* is a pair (f, g). Here are some examples:

$$f(x) = -\lfloor x \rfloor, \quad g(x) = \begin{cases} x - \lfloor x \rfloor & x \geq 0, \\ x, & x \leq 0, \end{cases} \tag{2}$$

$$f(x) = -\lfloor x \rfloor, \quad g(x) = \begin{cases} x - \lfloor x \rfloor & x \leq 0, \\ x, & x \geq 0 \end{cases} \tag{3}$$

(note the lack of uniqueness in g, for a given f); or

$$f : \mathbf{R} \to \{0, 1\} \text{ arbitrary with } f(x + 1) = f(x); \quad g(x) = x + f(x). \tag{4}$$

There are many others. Verification that these pairs indeed satisfy (1) is straightforward. The function f in (4) is, in fact, the characteristic function of some set $E \subset [0,1)$, extended by periodicity to the whole real line. Since E can be *any* set in the unit interval, we already get a family of solutions,

109

equipotent with the class of all functions $\mathbf{R} \to \mathbf{R}$. Taking E to be the empty set, we find the (trivial) pair

$$f(x) = 0, \quad g(x) = x. \tag{5}$$

These examples can be further modified. For example, in (4), instead of a two-valued function f, one might take any integer-valued, 1-periodic function f. Another useful observation is the following. Let $f, g : \mathbf{R} \to \mathbf{R}$ be any pair of functions and $b, c \in \mathbf{R}$ be any pair of real numbers, and define

$$\bar{f}(x) = f(x) + b, \quad \bar{g}(x) = g(x - b) + c. \tag{6}$$

Then

$$f, g \text{ satisfy (1) if and only if } \bar{f}, \bar{g} \text{ satisfy (1)}. \tag{7}$$

This is so because the value at (x, y) of the left-hand side of (1) for f, g and the value of the left-hand side of (1) for $\bar{f}, \bar{g}$ differ by a constant:

$$x + \bar{g}(y + \bar{f}(x)) = x + g(y + \bar{f}(x) - b) + c = x + g(y + f(x)) + c,$$

and the same applies to the right-hand side of (1).

In view of (6) and (7), there is no loss of generality in restricting attention to the case where

$$f(0) = g(0) = 0, \tag{8}$$

as any solution of the equation (a pair f, g) can be reduced to a solution with property (8), via a transformation of the type described in (6).

A general feature of functional equations in two (or more) independent variables is that symmetry impairs information. Here, equation (1) is pure symmetry; the expression on the right is the mirror image of that on the left, after interchanging x and y. The existence of a vast variety of solutions is therefore no surprise. Note, however, that all of the examples (2)–(4) are discontinuous functions, with the lone exception of (5).

Solutions showing much more regularity can also be given. We exhibit three important families:

$$f(x) = \frac{1}{r} \cdot \ln(ae^{-rx} + 1 - a),$$

$$g(x) = \frac{1}{r} \cdot \ln\left(\frac{e^{rx} + a}{1 + a}\right) \quad (r \neq 0; 0 \leq a \leq 1), \tag{9}$$

$$f(x) = -ax, \quad g(x) = \frac{x}{1+a} \quad (a \neq -1), \tag{10}$$

$$f(x) = -x, \quad g(x) = \frac{x}{2} + h(x); \quad h : \mathbf{R} \to \mathbf{R} \text{ any even function.} \tag{11}$$

Again, verification that these actually are solutions is a matter of routine calculation.

Now, (9) is a two-parameter family of pairs of functions, (10) is a one-parameter family, and they each satisfy $f(0) = g(0) = 0$. Dropping this last condition and applying transformation (6), we enlarge these families by two more degrees of freedom. Family (11) is more puzzling — it shows a lot of freedom, involving an *arbitrary* even function h, which of course can be as regular or irregular as we please.

These three families only scarcely overlap: (9) coincides with (10) only for $a = 0$ (becoming (5)); and each of (9), (10) coincides with an instance of (11) for $a = 1$ (and a suitable even function h).

2. Means

Means are mentioned in the title of this note. Everybody knows that if we consider, say, the arithmetic mean of N numbers, and if we select some n of them $(n < N)$ and replace them by *their* arithmetic mean (repeated n times), the overall average shall not change. The same happens with the geometric mean of N positive numbers, or any other power mean. I call this feature *compatibility*. To be precise:

A *mean* is defined here as any function

$$\mu : \mathbf{R}_+^n \to \mathbf{R}_+; \quad \mathbf{R}_+ = (0, \infty), \quad n \in \mathbf{N}, n \geq 2, \tag{12}$$

which is symmetric, with $\mu(\mathbf{x}) = \mu(\mathbf{x}')$ for any permutation $\mathbf{x}'$ of $\mathbf{x} = (x_1, \ldots, x_n)$, non-decreasing in each x_i, and normalized, with $\mu(t, \ldots, t) = t$ for $t \in \mathbf{R}_+$. It is *homogeneous* if $\mu(t\mathbf{x}) = t\mu(\mathbf{x})$ for $t \in \mathbf{R}_+$, $\mathbf{x} \in \mathbf{R}_+^n$. The *power means* with exponent $r \in \mathbf{R}$, defined by

$$\left(\frac{1}{n} \sum_{i=1}^n x_i^r \right)^{1/r} \text{ for } r \neq 0 \text{ and } \left(\prod_{i=1}^n x_i \right)^{1/n} \text{ for } r = 0, \tag{13}$$

are a model example.

Now, suppose that μ is some homogeneous mean in $\mathbf{R}_+^n$ and $\tilde{\mu}$ is some homogeneous mean in $\mathbf{R}_+^{n+1}$. We say that μ and $\tilde{\mu}$ are *compatible* if, for any $(x_1, \ldots, x_{n+1}) \in \mathbf{R}_+^{n+1}$

$$\tilde{\mu}(x_1, \ldots, x_n, x_{n+1}) = \tilde{\mu}\left(\underbrace{z, \ldots, z}_{n}, x_{n+1}\right) \text{ where } z = \mu(x_1, \ldots, x_n).$$

$$(14)$$

The power means (in $\mathbf{R}_+^n$, $\mathbf{R}_+^{n+1}$ with the same exponent r) satisfy this property. This is basically due to the fact that their defining formula (13) has the same shape and is equally meaningful, both in n variables and in $n+1$ variables. What about other examples (with μ and $\tilde{\mu}$ not necessarily given by any algebraic formula)? A tool to address this question is the following observation.

Let μ and $\tilde{\mu}$ be compatible homogeneous means in $\mathbf{R}_+^n$ and $\mathbf{R}_+^{n+1}$. Define

$$f(x) = \ln \mu\left(\underbrace{1, \ldots, 1}_{n-1}, e^{-x}\right),$$

$$g(x) = \ln \tilde{\mu}\left(\underbrace{e^x, \ldots, e^x}_{n}, 1\right) \quad \text{for } x \in \mathbf{R}, \tag{15}$$

and call $f, g : \mathbf{R} \to \mathbf{R}$ the functions *induced* by μ and $\tilde{\mu}$. Clearly, we have $f(0) = g(0) = 0$, f is non-increasing and g is non-decreasing. We now show that f and g satisfy equation (1).

Choose $x, y \in \mathbf{R}$ and consider the number

$$M(x, y) = \tilde{\mu}\left(\underbrace{1, \ldots, 1}_{n-1}, e^{-x}, e^{-y}\right).$$

Writing

$$z = e^{f(x)} = \mu(\underbrace{1, \ldots, 1}_{n-1}, e^{-x}) \quad \text{and} \quad w = ze^y = e^{y+f(x)},$$

we obtain by compatibility and homogeneity,

$$M(x, y) = \tilde{\mu}\left(\underbrace{z, \ldots, z}_{n}, e^{-y}\right) = e^{-y}\tilde{\mu}\left(\underbrace{w, \ldots, w}_{n}, 1\right).$$

Taking logarithms, (15) and $w = e^{y+f(x)}$ yield

$$\ln M(x, y) = -y + g(y + f(x)).$$

In the last equation, the left side is symmetric in x and y, and we therefore obtain

$$x + g(y + f(x)) = y + g(x + f(y)),$$

and this is equation (1).

That's how equation (1) came into being. It can be easily checked that the rth power means (13), with the same exponent $r \neq 0$, induce via (15) the pairs f, g of type (9) with $a = 1/n$, and the geometric means (13), $r = 0$, induce the pair (10), also with $a = 1/n$.

Less evident is that if *some* pair μ and $\tilde{\mu}$ (means in $\mathbf{R}_+^n$, $\mathbf{R}_+^{n+1}$) induce via (15) a pair f, g of type (9) or (10), with parameters a and r, then automatically $a = 1/n$; a proof is given in M.E. Kuczma (1993).

On the other hand, the third family — pairs f, g of type (11) — never occur as functions induced by compatible homogeneous means. This is true because (see (11)), if $f(x) = -x$ then the formula in (15) would imply $e^{-x} = \mu(1, \dots, 1, e^{-x})$ for all $x \in \mathbf{R}$, whence by setting $x = -1$ and $x = 1$ (and by coordinatewise monotonicity) we would get

$$e = \mu(1, 1, \dots, 1, e) \leq \mu(1, e, \dots, e, e) = e\mu(e^{-1}, 1, \dots, 1, 1)$$

$$= e \cdot e^{-1} = 1,$$

which is a contradiction.

3. Solutions of the Functional Equation

Taking all of this into account, we see that if, *in some function class F*, equation (1) with condition (8) has no other solutions than those of types (9), (10), (11), then for each $n \in \mathbf{N}$, $n \geq 2$, there are no pairs of homogeneous compatible means (one in $\mathbf{R}_+^n$ and the other in $\mathbf{R}_+^{n+1}$), other than power means (13), that would coordinate-wise be functions of class F.

So, what regularity conditions on functions f, g satisfying (1) with (8) guarantee that they must belong to type (9), (10) or (11)?

My first step in this direction was to show that analyticity of both f and g suffices (or analyticity of one of them plus some technical condition on the other one); basically, that was work on power series (Kuczma, 1993). Closer inspection of their low-order coefficients led to the hypothesis that second-order differentiability of f, g should also suffice. This was turned into a theorem by Justyna Sikorska (1998). Shortly later, Nicole Brillouët-Belluot (2004) showed that first-order differentiability is enough.

A further significant push was made again by Sikorska (2003): if f, g, satisfying (1) with (8), are continuous, strictly monotone, and convex or concave, then they are of type (9), (10) or (11). Accordingly, if μ and $\tilde{\mu}$ are compatible homogeneous means, of coordinate-wise regularity as above, they must be power means.

Sikorska (2003), in the same paper, obtained yet another interesting result, which is more easily stated assuming that f, g satisfy equation (1) and dropping condition (8): if f, g are continuous, monotone, but not strictly monotone, and f is not identically zero, then they are of one of the two following types (for some $a, b, c \in \mathbf{R}$):

$$f(x) = \max\{b, a + b - x\}, \quad g(x) = \max\{c, x + c - a - b\} \quad (16)$$

or

$$f(x) = \min\{b, a + b - x\}, \quad g(x) = \min\{c, x + c - a - b\}. \quad (17)$$

It is worth noting that (16) with $a = b = c = 0$ represents the pair of functions induced by the means $\max_{i \leq n} x_i$, $\max_{i \leq n+1} x_i$ and analogous is the claim about the pair (17), with min in place max. (Of course, these are the limit cases of the power means (13) as $r \to \infty$, resp. $r \to -\infty$.)

Although the equation has arisen merely as a tool in research on means and their compatibility, it apparently attracted more attention than the parent problem and started a life of its own. Note, however, that the results just quoted were obtained some considerable time ago. I have not heard about any progress since then. The major problem still waiting for an answer is:

Find the general solution of equation (1) in the class of continuous functions.

In other words:

Are there any continuous functions satisfying equation (1) with condition (8) other than those pertaining to families (9), (10), (11), (16), (17)?

If not, that shall mean that there are no pairs of compatible homogeneous means, one in $\mathbf{R}_+^n$, the other in $\mathbf{R}_+^{n+1}$, other than the power means (13) or their limit max/min forms. This holds because the conditions adopted as the definition of a homogeneous mean (see the text around (12)) imply its continuity (a rather standard exercise in multivariate calculus). Consequently, pairs of functions f, g induced by pairs of means as above are automatically continuous.

Note also that the question of the general solution of (1) in the class of *monotone* functions has so far found no satisfactory answer.

4. Conclusion

What is the rationale for a talk like this at a WFNMC conference (or including this chapter in a book of competition problems, for that matter), other than merely reporting a piece of rather specialized research? My intention was, of course, to give some publicity to equation (1) — a nice functional equation, after all. But not only that. The presentation was primarily addressed to listeners and readers who are teachers, instructors, and coaches of olympiad participants. The open problem of finding a full solution of (1) in continuous functions (or monotone functions) can be attractive to the young. It can be hard, but is not likely to require advanced knowledge or techniques. I believe that it might be solved by some young student, and it seems more important that he or she be smart, rather than particularly experienced.

The word *problem* has different meanings in our community. An *open* problem is one of them. But we also use this term when preparing a classroom test, an exam or a competition, as a substitute for *exercise* (with a solution known to the proposer), perhaps more challenging as compared to everyday classroom work. So, in conclusion, I wish to propose, as a Math-Comp-bonus, an olympiad-style "problem" related to the topics discussed above. The function F which appears in it obviously imitates the $\tilde{\mu}$ of the foregoing considerations (and is visibly compatible with the root mean square in $\mathbf{R}_+^2$).

Let $F : \mathbf{R}_+^3 \to \mathbf{R}_+$ satisfy

$$F(x, y, z) = F\left(\sqrt{\frac{x^2 + y^2}{2}}, \sqrt{\frac{x^2 + y^2}{2}}, z\right) = F(y, z, x) = F(z, x, y).$$

Show that if $a^2 + b^2 + c^2 = u^2 + v^2 + w^2$, then $F(a, b, c) = F(u, v, w)$.

(Note that the conditions say nothing about continuity.)

The exercise admits several possible variations: $\mathbf{R}_+^n$ in place of $\mathbf{R}_+^3$, averaging not in pairs but rather in triples or other k-tuples; the root mean square (of x, y or, in an extended version, of a k-tuple) replaced by some other power mean. If one starts to seek far reaching generalizations, one will probably sooner or later arrive at the actual problems that comprise the subject of this talk.

And that's the end of the story for today.

Bibliography

Brillouët-Belluot, N. (2004). On a symmetric functional equation in two variables, *Aequationes Math.* 68, 10–20.

Kuczma, M.E. (1993). On the mutual noncompatibility of homogeneous analytic non-power means, *Aequationes Math.* 45, 300–321.

Sikorska, J. (1998). Differentiable solutions of a functional equation related to the non-power means, *Aequationes Math.* 55, 146–152.

Sikorska, J. (2003). On a functional equation related to power means, *Aequationes Math.* 66, 261–276.

Chapter 1.7

Open Problems as Generalizations of Tasks from Mathematics Competitions

Kiril Bankov

University of Sofia and Bulgarian Academy of Sciences, Bulgaria

1. Introduction

There are hundreds of mathematics competitions around the world. Thousands of students participate in them and win awards every year. The names of the most successful among them find their way into the lists of results, building the history of the competitions. No doubt, these lists include an impressive number of people, many of whom go on to careers as professional mathematicians. This is one of the goals of mathematics competitions: to stimulate the development of mathematical talents.

Of course, it is not only the participants that build the history of competitions. In fact, for any competition, there are two fundamentally important sets: the set of participants, and the set of problems. The problems constitute the intellectual product that survives in the history. In the course of time, hundreds of thousands of fascinating ideas have been implemented in the problems presented in mathematics competitions, and it is impossible to review or consider all of them. There are, however, two types of "unforgettable" problems that deserve special attention:

(i) Problems that do not require specific knowledge from the mathematics school curriculum in order to be understood, but whose solutions require deep thinking, mathematical reasoning, experience, and a lot of intuition. Sometimes their solutions are quite unexpected. These problems are some of the best examples of the beauty of mathematics.

(ii) Problems that lead to interesting generalizations. Sometimes these can be sources of interesting unsolved problems, able to actively stimulate mathematical research.

This chapter presents some examples of such unsolved problems inspired by problems presented at mathematics competitions.

2. Exploring A Period

We consider the following situation. Let $n > 2$ cells be arranged in a circle. Each cell is initially occupied by either 1 or 0. The following operation is then performed on the arrangement of cells: First, we introduce n new cells, one between any two of the existing ones. We then write numbers into the new cells; 0 if the numbers in the original neighboring cells are equal, and 1 if these numbers are different. We then delete the original cells, which yields a new arrangement of n cells.

This situation was used in a problem presented at a mathematics competition in 1975 in the former Yugoslavia.

Problem 2.1. In the given situation, let $n = 9$ and let four of the cells be occupied by 1, and the other five be occupied by 0. Is there an arrangement such that it is possible to obtain zeros in all nine cells in a finite number of admissible steps?

Solution. The answer is no, and here is the argument. Assume that all nine cells contain zeros after a finite number of admissible steps. Then all nine cells contain ones in the arrangement next to last. Therefore, in the arrangement before that, any two neighboring cells contain different numbers, which is impossible since there is an odd number of cells.

This problem gives rise to a variety of generalizations. Different variations of the initial arrangements can be considered, depending on the number n of the cells, and on the number and the positions of the initial ones and zeros. One interesting generalization is presented in the following problem.

Problem 2.2. In the given situation, there is a 1 in one cell initially and all zeros elsewhere. For which values of n is it possible to obtain zeros in all cells in a finite number of admissible steps?

Let n be a number for which it is possible to obtain zeros in all cells in a finite number of steps. Because of the arguments presented in the solution of Problem 2.1, n must be even, and we can limit our discussion to even values of n.

I originally presented this as an open problem in my talk at Topic Study Group 30 at the International Congress on Mathematical Education (ICME-13) in Hamburg in 2016. At the time I did not know the answer. Soon after the congress, I was preparing my talk as a chapter for a book, and attempted to find a complete solution to the problem in order to be able to include it. My desk was full of sheets of papers with circles and numbers. I used a computer to check for different values of n. Time was passing, and just before the submission deadline, I realized that Sierpinski's triangle could help. With the aid of this tool, I was then able to prove that the answer to Problem 2.2 is that the required values of n are precisely the powers of 2, and the proof was submitted for publication just in time (Bankov, 2017).

Having solved Problem 2.2, the next question is, what happens for the other values of n.

If n is not a power of 2, we now know that the execution of the operation will never end with the given structure. On the other hand, there are only finitely many arrangements of zeros and ones in the n cells, no matter which value of n we start with. This means that the arrangements of the numbers must repeat cyclically after a certain number of steps. If we call smallest number of steps in such a cycle a *period*, we can consider the following:

Open Problem. Explore how the period depends on n.

The following table presents the values of the periods for some values of n.

n	Period
10	8
11	32
13	64
14	16
19	512
23	2048
29	16,384
48	32
58	32,768

Taking a close look at the values in this table, we can conjecture that the period must always be a power of 2, but a proof of this is not yet available.

3. Large Subsets of Disjoint Figures

This section presents an open problem that results as a generalization of a problem posed at two famous mathematics competitions, concerning a certain set of disjoint figures in the plane. Before presenting the problems, let us consider the one-dimensional case.

Problem 3.1. Let M be a finite set of segments on the line, with the sum of their lengths equal to L. There then exists a disjoint subset of M, such that the sum of the lengths of the segments in M is not less than $\frac{L}{2}$.

A solution to this problem can be found in Shkliarskii *et al.* (1974). Mathematically, the more interesting result states that this statement presents the best possible result, i.e. for every $\delta > 0$ there is a covering of a given segment of length L by a finite set M of segments, such that the sum of the lengths of the segments of any disjoint subset is less than $\frac{L}{2} + \delta$.

A possible two-dimensional case is presented in the following problem, which was set at the Moscow Mathematical Olympiad 1979, and the Austrian–Polish Mathematics Competition 1983.

Problem 3.2. Let M be a finite set of circles in the plane, such that the area of the union of all circles is equal to A. Prove that there exists a disjoint subset of M, such that the sum of the areas of the elements of the subset is not less than $\frac{A}{9}$.

Solution. The proof uses the principle of mathematical induction on the number of circles in the set M. The statement is obvious if M contains 1 or 2 circles. Let n be a natural number, $n \geq 3$ and assume that the statement is true if M contains k circles for every $k < n$. We will prove that the statement is then also true if M contains n circles, $M = \{K_1, K_2, \ldots, K_n\}$.

Let K be the circle in M with the largest radius R. Let $A(K)$ denote the area of K. If $A(K) \geq \frac{A}{9}$, the required subset consists of one circle, namely K. Otherwise, let $3K$ denote the circle concentric with K, with radius $3R$. If a circle C of M has a common point with K, then C lies in the interior of $3K$, because R is the largest radius of the circles of M (Figure 3.1). Because we have assumed $A(3K) = 9A(K) < A$, there are therefore

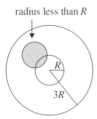

radius less than R

Figure 3.1. Illustration of the proof.

circles in M that do not have a common point with K. Let N denote the set of these circles. Obviously, the area of the union of the circles of N is not less than $A - 9A(K)$. According to the inductive assumption, there exists a disjoint subset P of N, such that the sum of the areas of the circles in N is not less than $\frac{A-9A(K)}{9} = \frac{A}{9} - A(K)$. The set $P \setminus \{K\}$ is then a disjoint subset of M, such that the sum of the areas of its elements is not less than $\frac{A}{9}$, as claimed.

I have presented this solution to Problem 3.2 because the same method can be used for the proof of a generalization of this problem. The generalization considers not only circles, but any set of bounded figures. In order to consider this problem, we first need to familiarize ourselves with the concept of a neighborhood.

Let $d(X, Y)$ denote the distance between points X and Y in the plane, and let K be a bounded figure in the plane. The number $d(K) = \sup\{d(X, Y); X, Y \in k\}$ is called the *diameter* of K. Let Z be a point in the same plane. We then call the number $d(Z, K) = \inf\{d(Z, Y); Y \in k\}$ the *distance* between the point Z and the figure K. For any $\varepsilon > 0$, the *neighborhood* $O_\varepsilon(K)$ of K with radius ε is then defined as the set of all points X in the plane whose distance from K is not greater than ε, i.e. $O_\varepsilon(K) = \{X; d(X, K) \leq \varepsilon\}$.

To visualize the notion of a neighborhood, consider the following thought experiment. Throw a figure K into a basin of water and look how the waves spread. Their shape then has the form of a neighborhood of K.

For example, if K is a circle with radius R, the neighborhood of K is a concentric circle with radius $R + \varepsilon$. The neighborhood of a square with side a is shown in Figure 3.2. Its area is equal to $a^2 + 4a\varepsilon + \pi\varepsilon^2$. The neighborhood of an equilateral triangle with side a is shown in Figure 3.3. Its area is equal to $\frac{a\sqrt{3}}{4} + 3a\varepsilon + \pi\varepsilon^2$.

Figure 3.2. Neighborhood of a square.

Figure 3.3. Neighborhood of an equilateral triangle.

We are now ready to consider the next theorem, which is a generalization of Problem 3.2.

Theorem. *Let* $M = \{K_1, K_2, \ldots, K_m\}$ *be a finite set of bounded figures in the plane, the union of which has an area equal to* A. *For every* $i = 1, 2, \ldots, m$, *let* d_i *denote the diameter of* K_i, $\lambda_i = \frac{A(K_i)}{A(O_{d_i}(K_i))}$, *and* $\lambda = \min\{\lambda_1, \lambda_2, \ldots, \lambda_m\}$. *Then there exists a disjoint subset of* M, *such that the sum of the areas of the elements of the subset is not less than* λA.

The proof of this theorem (Bankov, 1996) uses the method of the solution of Problem 3.2. In order to see this, let K be the element of M with the largest diameter. If $A(K) \geq \lambda A$, we are done. If not, consider the set N of figures in M that do not have a common point with K. First, prove that N is not empty. Next, apply the principle of Mathematical Induction to select a disjoint subset of N with a "large" area, and add K to this subset.

Corollary. *Let* M *be a finite set of squares in the plane, the union of which has an area equal to* A. *Then there exists a disjoint subset of* M, *such that the sum of the areas of the elements of the subset is greater than* $\frac{A}{13}$.

This follows from the fact that in the case of squares, we have

$$\lambda = \lambda_1 = \lambda_2 = \cdots = \lambda_m = \frac{1}{1 + 4\sqrt{2} + 2\pi} > \frac{1}{13}.$$

A similar result can easily be found for a finite set of equilateral triangles in the plane, and this is left as an exercise for the interested reader.

The question now is whether the numbers $\frac{1}{9}$ (in Problem 3.2), λ (in the theorem), and $\frac{1}{13}$ (in the Corollary) give the best possible results. In other words, is it true that increasing any of these numbers (for example λ) by any "small" number $\delta > 0$, means that there is certainly a set of bounded figures covering the area A such that any disjoint subset covers an area less than $\lambda + \delta$. The answer is negative. For example, in Shkliarskii *et al.* (1974), the following problem can be found.

Problem 3.3. A square K of area A is covered by a finite set M of squares, whose sides are parallel to the sides of K. Prove that there exists a disjoint subset of M, such that the sum of the areas of its elements is not less than $\frac{A}{9}$.

Certainly, Problem 3.3 yields a better result than can be obtained by the Theorem.

Here is what is known about the best possible result.

Let Ω be a class of bounded figures in the plane. For example, Ω could be the class of all squares with parallel sides, or the class of all squares, or the class of all circles, or the class of all regular polygons, etc. Let $M = \{K_1, K_2, \ldots, K_m; K_i \in \Omega, i = 1, 2, \ldots, m\}$ be a finite set of figures in the plane, each of which belongs to Ω. Let A denote the area of the union of the figures of M, $D(M) = \{K_{i_1}, K_{i_2}, \ldots, K_{i_n}\}$ a disjoint subset of M, and $A(D(M)) = A(K_{i_1}) + A(K_{i_2}) + \ldots + A(K_{i_n})$. The main problem is to find the number

$$\mu = \inf_{M \in \Omega} \sup_{D(M)} A(D(M)).$$

The notation $M \in \Omega$ means that M is any finite set of figures of Ω.

Roughly speaking, in order to determine μ, we can consider all disjoint subsets of M, take the one with the maximum area, and then consider all finite sets of figures of Ω and take the minimum across these maximum areas.

Certainly, the number μ depends on the class Ω. The determination of μ is not an easy task. Here are some results.

Let Ω be the class of all *squares with parallel sides*. The Hungarian mathematician T. Rado (1928) conjectured that $\mu = \frac{A}{4}$. The German mathematician R. Rado (1950) proved that $\mu > \frac{A}{8.75}$. Ten years later, the Russian mathematician Zalgaller (1960) proved that $\mu > \frac{A}{8.6}$. Until 1973 the conjecture of T. Rado seemed to be true, but the Hungarian mathematician Ajtai (1973) constructed a set of squares with parallel sides which disproves this. This made the problem much more attractive.

The next open problems focus on concrete sets Ω.

Open Problem 3.4. Find μ if Ω is the class of all squares with parallel sides.

Open Problem 3.5. Find μ if Ω is the class of all circles.

Open Problem 3.6. Find μ if Ω is the class of all squares.

Open Problem 3.7. Find μ if Ω is the class of all equilateral triangles.

Open Problem 3.8. Find μ if Ω is the class of all regular polygons.

Certainly, some other similar open problems arise, which the interested reader may be inspired to create themselves.

Bibliography

Ajtai, M. (1973). The solution of a problem of T. Rado, *Bull. Acad. Pol. Sci. Ser. Sci. Math. Astron. Et Phys.* 21, 61–63.

Bankov, K. (1996). Selection of a large subset of disjoint figures, *Geombinatorics* VI(2), 41–47.

Bankov, K. (2017). Arrangements and transformations of numbers on a circle: An essay inspired by problems of mathematics competitions, in *Competitions for Young Mathematicians. Perspectives from Five Continents*, edited by A. Soifer. Springer, pp. 101–122.

Rado, R. (1949). Some covering theorems, *Proc. London Math. Soc.* 51, 232–264.

Rado, T. (1928). Sur une probleme relative a une teoreme de Vitali, *Fundamenta Math.* 11, 228–229.

Shkliarskii, D. O., Chentsov, N. N., and Yaglom, I. M. (1974). Geometricheskie otsenki i zadachi iz kombinatornoi geometrii [*Geometric Estimates and Problems from Combinatorial Geometry*] (in Russian). Nauka: Moscow, pp. 203–204.

Zalgaller, V. (1960). A note of Rado's theorem, *Matematicheskoe Prosveschenie* 5, 141–148 (in Russian).

Part 2

Some Favorite Puzzles and Problems Presented by Participants

Chapter 2.1

Introduction, Problems and Solutions

1. Introduction

As its name unsubtly suggests, the World Federation of National Mathematics Competitions is a collection of people from all over the world whose driving common interest is the organization of mathematics competitions. This of course implies a strong interest in the content of such activities, i.e. mathematical puzzles and problems. These can range in scope and difficulty level from bite-sized quiz questions as you might find in your daily newspaper right up to advanced Olympiad-style problems that aren't really all that far away from serious mathematical research.

In this special part, we have collected some of the participants' favorite problems, as presented at the congress. Most of these were discussed during the open problem sessions of the congress, but some had been prepared for participation and then had to be shelved for lack of time. It is hoped that they will all find an interested audience among the readers of this book. Also included are some of the mathematical puzzles that the participants posed to each other at the congress excursion dinner (when this group gets together, it seems you can't get them off topic for long ...), along with a special problem from one of the talks that caught the attention of several congress participants in a particularly inspiring way.

All problems presented in this part are original creations of the respective presenters, as far as is known, unless specifically stated otherwise. Since the ideas for such problems are often "in the air", it may well be that some such a problem has already been posed elsewhere, and if so, we offer our apologies to the unknown problem author(s). In order for readers to be able to try their hands at solving these puzzlers themselves, all problems are first posed in Section 2, with their respective solutions presented in Section 3. We start with some simple but ingenious mathematical puzzles and things then become successively more abstract (and possibly more difficult), finishing with some quite challenging competition-style questions.

We hope that you will enjoy thinking about these problems as much as the congress participants did.

2. Problems

2.1. *Problems suggested by Dima Nikolenkov (Switzerland)*

The following number puzzles kept us busy all during the excursion dinner in Graz.

Problem 2.1. Change one dot (one pixel) in order to create a mathematically true statement.

As usual with such things, it is tough to trace the origin of this puzzle. On the internet, this problem can be found at https://www.redd it.com/r/mathpuzzles/comments/88fcb2/one_of_these_pixels_on_the_sco reboard_should_be/.

Problem 2.2. Add exactly one line segment to make a valid equation

$$5 + 5 + 5 + 5 = 555$$

As with Problem 2.1, the origin of this is unclear, but it too can be found on reddit.

Problem 2.3. Move exactly one digit to make a valid equation

$$63 + 1 = 62$$

This is similar to a puzzle by Mel Stover on puzzles.com.

Problem 2.4. Move exactly one digit to make a valid equation

$$101 - 102 = 1$$

I first saw this one in E. G. Kozlova's book *Skazki i podskazki*, Moskow; http://ilib.mccme.ru/pdf/kozlova.pdf.

Problem 2.5. A specific time on a specific day of the year is called *pandigital* if it is written as

$$DD : MM : HH : MM : SS,$$

where the pairs of digits represent the day, month, hour, minute and second respectively, including leading zeroes and all digits from 0 to 9 are used exactly once. What are the earliest and latest pandigital times in the year?

This was published in Puzzleart, Tokyo 1992 and can also be found in Nobuyuki Yoshigahara, *Puzzles 101, A Puzzlemaster's Challenge*, AK Peters, Natick, MA, 2004.

Problem 2.6.

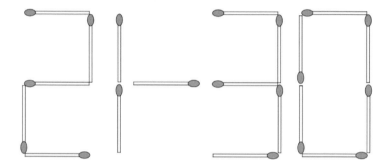

(a) (Warm-up) Move one match to get an expression with the value 1.
(b) Move one match to get an expression with the value 2018.

I think I saw something similar years ago in *Kvant* magazine. I used the idea to get the congress year as one of the answers.

2.2. *Problems suggested by Andy Liu (Canada)*

Problem 2.7 is a puzzle-type problem with somewhat unexpected solutions. Problems 2.8–2.11 are from Andy's upcoming book of problems from the Leningrad Mathematical Olympiad, and were originally posed at that competition in 1992.

Problem 2.7. The diagram shows six chisel-shaped pieces forming a rectangle whose width is three times that of each chisel. It is trivial to use these pieces to form a rectangle, whose width is equal to that of an individual chisel. Form a rectangle whose width is twice that of each individual chisel. The pieces may be rotated or reflected. There are three solutions.

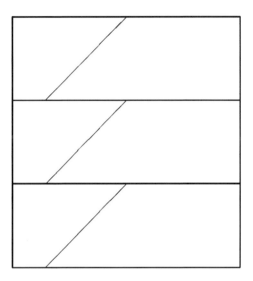

Problem 2.8. Anna and Boris play a game with 1992 counters. Anna starts, and turns alternate thereafter. In her first move, Anna can take any number of counters, but not all of them. In each subsequent move, a player can take any number of counters as long as that number is a divisor of the number of counters that the other player has just taken. The winner is the person who takes the last counter. Which player has a winning strategy?

Problem 2.9. There are 100 cities in a country, some pairs of which are connected by roads. If any city closes all the roads going in or out of it, one can still go from any city to any other except that city. Prove that the country can be divided into two provinces with 50 cities each so that in both provinces it is possible to go from any city to any other city.

Problem 2.10. A circle is divided into n sectors occupied by $n + 1$ frogs in a random distribution. In each move, two frogs in the same sector jump to adjacent sectors in different directions. Prove that after a finite number of moves, more than half the sectors are occupied by frogs.

Problem 2.11. Fyodor collects coins. No coin in his collection is more than 10 cm in diameter. He keeps all the coins arranged side by side in a rectangular box of size 30 cm × 70 cm. He gets a new coin of diameter 25 cm. Prove that he can fit all of his coins in a square box of side length 55 cm.

2.3. *Problems suggested by Iliana Tsvetkova (Bulgaria)*

The problems suggested by Iliana were used in various Bulgarian mathematics competitions for young students.

Problem 2.12. A digital clock shows the time as 21:06. How many minutes will it take until the clock shows the same four digits (in some different order) the next time? (High School of Mathematics in Sofia – exam 2016, 4th grade.)

Problem 2.13. A functioning calculator shows digits in the following way:

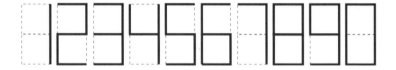

A broken calculator shows only horizontal traces. Determine the sum of all the digits in the expression, as shown by the broken calculator in the figure.

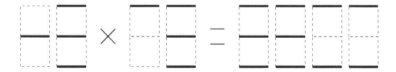

(First Private Mathematical School – Math Competition – 2013, 4th grade.)

Problem 2.14. In a triangle ABC, we are given $AC = BC$, $\angle ABM = 60°$, $\angle ACB = 20°$ and $\angle BAN = 50°$. Determine the size of $\angle AMN$. (National Math Competition – Pazardjik 1998, 7th grade.)

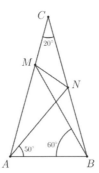

2.4. *Problems suggested by Mike Clapper (Australia)*

The problems suggested by Mike were taken from the AIMO (Australian Intermediate Mathematics Olympiad) 2014 and 2015.

Problem 2.15. Justin's sock drawer contains only identical black socks and identical white socks, a total of less than 50 socks altogether. If he withdraws two socks at random, the probability that he gets a pair of the same color is 0.5. What is the largest number of black socks he can have in his drawer?

Problem 2.16. X is a point inside an equilateral triangle ABC. Y is the foot of the perpendicular from X to AC, Z is the foot of the perpendicular from X to AB, and W is the foot of the perpendicular from X to BC. The ratio of the distances of X from the three sides of the triangle is $1:2:4$ as shown in the diagram. If the area of $AZXY$ is $13\,\text{cm}^2$, find the area of ABC.

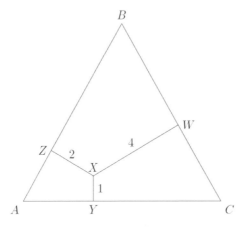

2.5. *Problem suggested by Romualdas Kašuba (Lithuania)*

The problem suggested by Romualdas was proposed by the Irish problem composer and long-term Irish IMO deputy leader Gordon Lessels. It was used as a problem in the Lithuanian team-contest 2016.

Problem 2.17. Nine stars ********* are written on a blackboard. John and Mary alternate turns, replacing them with the digits 1, 2, 3, 4, 5, 6, 7, 8, and 9. John, who starts the game, can replace any one of the stars with any one of the digits. When it is Mary's turn, she replaces any two of the remaining stars with any two of the remaining digits, and so on. After three moves, there is a 9-digit number on the blackboard, formed with each of the digits used exactly once. Mary wins if this resulting 9-digit number is divisible by 27. If it is not, John wins. Can Mary always win the game, independent of John's moves? What strategy must she apply?

2.6. *Problems suggested by Erich Fuchs and Bettina Kreuzer (Germany)*

The problems suggested by Erich and Bettina are from the environment of the Náboj competition.

Problem 2.18. Thirteen bees, one little bee and twelve large bees, are living in a 37-cell honeycomb. Each large bee occupies three pairwise adjacent

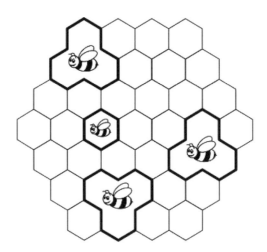

cells and the little bee occupies exactly one cell, as in the picture. In how many ways can the honeycomb be divided into thirteen non-overlapping sectors so that all thirteen bees can be accommodated in accordance with the given restrictions? (Náboj 2016, Problem 39, suggested by Łukasz Bożyk).

Problem 2.19. We are given a quadrilateral $ABCD$ with side lengths as shown in the picture. How long is $CF + FD$ if we know that $BE + EC = 78$ holds? (suggested by Erich Fuchs)

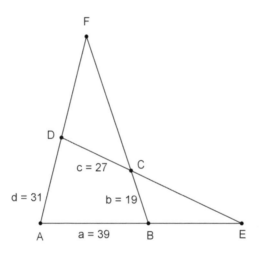

Problem 2.20. Gilbert Bates, a very rich man, wants to have a new swimming pool installed in his garden. Since he likes symmetry, he tells the gardener to build an elliptical pool within a $10\,\text{m} \times 10\,\text{m}$ square $ABCD$. This elliptical pool is to touch all four sides of the square, and particularly the side AB at a point P which is $2.5\,\text{m}$ from A. The gardener, who knows how to construct an ellipse if the foci and a point on the ellipse are given, realizes that, due to symmetry, he only needs to determine the distance between the foci in order to construct the ellipse. Can you help him by computing the distance between the foci in meters? (Náboj 2015, Problem 53, suggested by Erich Fuchs).

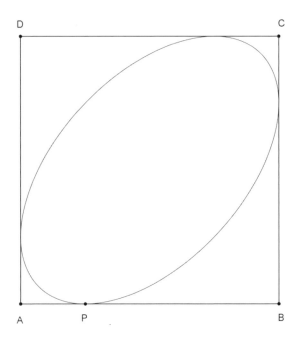

2.7. *Problems suggested by Alexander Soifer (USA)*

Here are some favorite problems of the federation's outgoing president.

Problem 2.21. Forty-one rooks are placed on a 10×10-chessboard. Prove that you can certainly choose five of them that do not attack each other. (We say that two rooks "attack" each other if they are in the same row or column of the chessboard.) (Proposed by Soifer and Slobodnik in 1972; published in Kvant in 1973. The authors were paid a whopping 40 rubles.)

Problem 2.22. (a) Each member of two 7-member chess teams is to play once against each member of the opposing team. Prove that as soon as 22 games have been played, we can choose 4 players and seat them at a round table so that each pair of neighbors has already played.

(b) Prove that 22 is the best possible value, i.e., the result cannot be guaranteed after 21 games. (Soifer, 2003, Bavarian Alps)

2.8. *Problems suggested by Marcin E. Kuczma (Poland)*

The following four problems are "quickies" proposed by Marcin. Each of them has a surprising and amusing twist to it.

Problem 2.23 — Intriguing Fifties. Let S be a 5-element set. How many functions $f : S \to S$ are there, such that $f^{50}(x) = x$ for each $x \in S$? Here $f^{50} = f \circ \cdots \circ f$ (50th iterate). (This problem was created for the 50th Mathematical Olympiad in Poland.)   ☐

Problem 2.24 — Exceed Galaxy Size. Imagine an $n \times n \times n$ cube as a loose structure composed of n^3 unit cubes. The task is to split it into n^3 single cubes. This of course can be done with one efficient kick. Ignoring this kind of triviality, we introduce a rule to be observed: at any one time it is allowed to remove any single small cube, whose upper face is visible from above. The decomposition task can be completed in many ways (really many; such combinatorial quantities like to grow rapidly...).

So, what is the smallest n for which the number of ways exceeds the diameter of our galaxy, measured in kilometers?

Suggestion: Before looking for an explicit formula, reflect for a while and try to come up with a reasonable guess (which can be fun; $n = 1$ is not enough, $n = 100$ is more than enough). By the way, according to Wikipedia, the diameter of the galaxy is approximated to be $(2 \pm 1) \cdot 10^5$ light years.

Problem 2.25 — A Swelling Ball. In a square with side-length 2, you can place four circles of diameter 1 in the corners, with each of them touching two sides of the square and two neighboring circles. In the center there is some space left for a tiny circle, touching those four. Similarly, in a cube with edge-length 2, one can accommodate eight balls of diameter 1 in the corners, with each touching three faces and three other balls; and a small ball, concentric with the cube, tangent to the eight diameter-1 balls. The same can be done in an n-dimensional cube of edge-length 2, with 2^n diameter-1 n-dimensional balls (hyperspheres) in the corners and another ball, of radius r_n, fitting in the central region and tangent to those 2^n ones. It is tempting to conjecture that $r_n \to 1$ as $n \to \infty$. Prove or disprove this.

Problem 2.26 — A Rapidly Increasing Sequence. Consider the rapidly growing sequence $a_n = \frac{n!}{n+1}$ and define $c_n = \prod_{k=0}^{n} a_k 2^{n-k}$

(for $n = 0, 1, 2, \ldots$). How many terms of the sequence $c_0, c_1, c_2, \ldots$ are numbers not greater than 2018? Try to guess, and then try to prove.

2.9. *Problem suggested by Walther Janous* (*Austria*)

This problem was perhaps the most talked about in the aftermath of the problem session at the congress.

Problem 2.27. A positive integer n is called *fantastic* if there exist positive rational numbers a and b such that $n = a + \frac{1}{a} + b + \frac{1}{b}$.

- (a) Prove that there exist infinitely many prime numbers p such that no multiple of p is fantastic.
- (b) Prove that there exist infinitely many prime numbers p such that some multiple of p is fantastic.

2.10. *Problem suggested by Hidetoshi Fukagawa* (*Japan*)

This problem was not presented as part of the problem sessions, but as part of Hidetoshi's talk on traditional Japanese Sangaku. His request for a nice solution from the audience inspired the wonderful solutions by Walther Janous and Marcin E. Kuczma presented in pages 166–170.

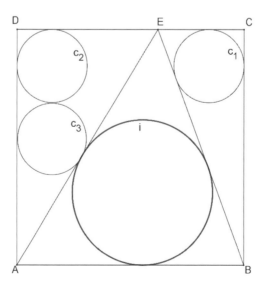

Problem 2.28. Three circles of equal radii are situated inside square $ABCD$. The first circle is tangent to sides BC and CD. The second circle, disjoint from the first, touches sides CD and DA. The third circle touches the second one and the side DA. The line through A, tangent to the third circle, and the line through B, tangent to the first circle, meet at a point E on side CD. Prove that the inradius of triangle ABE is twice as large as the radius of the three former circles. (traditional Sangaku)

2.11. *Problems suggested by Carlos Moreira (Brazil)*

The last two problems of this section were originally set at the fifth Iberoamerican University Mathematics Olympiad and the 34th Brazilian mathematical Olympiad, respectively.

Problem 2.29. Prove that there exist continuous functions $a_1, a_2, a_3, \ldots :$ $[0, 1] \to (0, +\infty)$ such that

(i) $\sum_{n=1}^{\infty} a_n(t) < +\infty$, for all $t \in [0, 1]$.
(ii) For every sequence (b_n) of positive real numbers with $\sum_{n=1}^{\infty} b_n < +\infty$ there exists a $t \in [0, 1]$, such that $\lim\limits_{n \to +\infty} \frac{b_n}{a_n(t)} = 0$.

Problem 2.30. Determine the smallest positive integer n for which a positive integer k exists, such that the last 2012 digits in the decimal representation of n^k are equal to 1.

3. Solutions

Solution to Problem 2.1. Change one dot (one pixel) in order to create a mathematically true statement.

Answer. In the last 1, erase the second dot from the bottom, yielding

The left side is $(71 + 1)(71 - 1) = 72 \times 70 = 5040$, and the right side is $7! = 5040$.

Solution to Problem 2.2. Add exactly one line segment to make a valid equation:

$$5 + 5 + 5 + 5 = 555$$

As with Problem 2.1, the origin of this is unclear, but it too can be found on reddit.

Answer. Cross one of the pluses to make 545

$$5 + 545 + 5 = 555$$

Note that a line through the equals sign, changing it to an unequal sign does yield a true statement, but not a "valid equation" as required.

Solution to Problem 2.3. Move exactly one digit to make a valid equation

$$63 + 1 = 62$$

Answer. Move the six on the right-hand side to the exponent.

$$63 + 1 = 2^6$$

Solution to Problem 2.4. Move exactly one digit to make a valid equation

$$101 - 102 = 1$$

Answer. Move the 2 to the exponent.

$$101 - 10^2 = 1$$

Solution to Problem 2.5. A specific time on a specific day of the year is called *pandigital* if it is written as

$$DD : MM : HH : MM : SS,$$

where the pairs of digits represent the day, month, hour, minute and second respectively, including leading zeroes and all digits from 0 to 9 are used exactly once. What are the earliest and latest pandigital times in the year?

Answer. The earliest pandigital time is on 26.03 at 17:48:59, while the latest is on 28.09 at 17:56:43.

Solution to Problem 2.6.

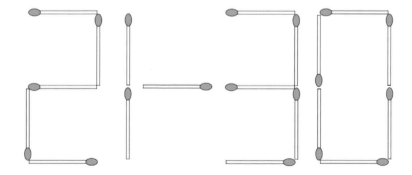

(a) (Warm-up) Move one match to get an expression with the value 1.

(b) Move one match to get an expression with the value 2018.

Answer. (a) Move either the vertical match at the bottom left in the two to make it a three or the other way around. This leads to $21 - 20 = 1$ or $31 - 30 = 1$.

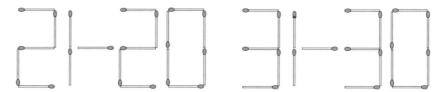

(b) Move the lower match of the one next to the higher one to make the exponent 11. This yields $2^{11} - 30 = 2048 - 30 = 2018$.

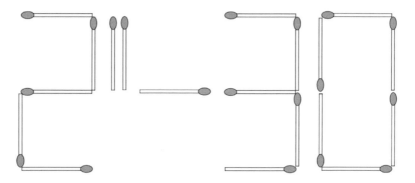

Solution to Problem 2.7. The diagram shows six chisel-shaped pieces forming a rectangle whose width is three times that of each chisel. It is trivial to use these pieces to form a rectangle, whose width is equal to that of an individual chisel. Form a rectangle whose width is twice that of each individual chisel. The pieces may be rotated or reflected. There are three solutions.

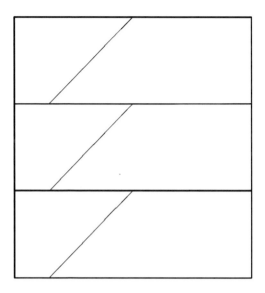

Answer. Here are the three solutions:

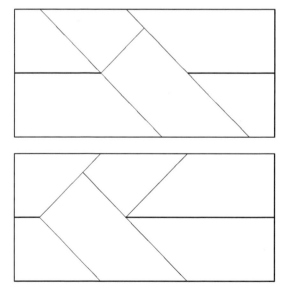

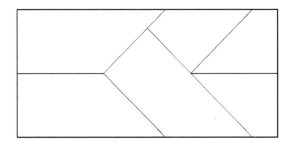

Solution to Problem 2.8. Anna and Boris play a game with 1992 counters. Anna starts, and turns alternate thereafter. In her first move, Anna can take any number of counters, but not all of them. In each subsequent move, a player can take any number of counters as long as that number is a divisor of the number of counters that the other player has just taken. The winner is the person who takes the last counter. Which player has a winning strategy?

Answer. Anna has a winning strategy, by taking 8 counters in her opening move. There are then $1984 = 64 \times 31$ counters left. After that, Anna simply copies the moves of Boris. Since the rest of the moves involve taking 1, 2, 4 or 8 counters and 1984 is divisible by 16, there will be an even number of moves after Anna's first move. Hence, she will make the last move and win.

Solution to Problem 2.9. There are 100 cities in a country, some pairs of which are connected by roads. If any city closes all the roads going in or out of it, one can still go from any city to any other except that city. Prove that the country can be divided into two provinces with 50 cities each so that in both provinces it is possible to go from any city to any other city.

Answer. We use a graph model, where vertices are cities and edges are roads. We shall prove by mathematical induction on k that the country can be divided satisfactorily into two provinces with k and $100 - k$ cities, respectively.

The case $k = 1$ follows from the given condition. Suppose we have a partition of the graph G into two induced subgraphs, each of which is connected. Let there be $k < 50$ vertices in G_1 and $100 - k$ vertices in G_2. We claim that we can move one vertex from G_2 to G_1, resulting in two induced subgraphs which are connected. Choose a vertex V in G_2 which is connected to some vertex in G_1. The move will not cause any problem for the enlarged G_1. Suppose what is left of G_2 has at least two components. Choose V so that the size of the smallest component C is minimal. Since $G\backslash\{V\}$ is connected, some vertex U in C is connected to some vertex in G_1. Suppose we move U over instead of V and what is left of G_2 has at least two components. Then one of them will be smaller than C, contrary to the minimality assumption on V. This contradiction justifies our claim.

Solution to Problem 2.10. A circle is divided into n sectors occupied by $n + 1$ frogs in a random distribution. In each move, two frogs in the same sector jump to adjacent sectors in different directions. Prove that after a finite number of moves, more than half the sectors are occupied by frogs.

Answer. Since there are more frogs than sectors, movement never ceases. We claim that every sector has been occupied at some point. Suppose there is a sector that has never been occupied. Number it n and the others from 1 to $n-1$ in cyclic order. Consider a move by two frogs from sector k, one going to sector $k - 1$ and one going to sector $k + 1$. The sum of the squares of the numbers of the sectors occupied by the $n + 1$ frogs increases by $(k - 1)^2 + (k+1)^2 - 2k^2 = 2$. This sum cannot increase without bound. So contrary to our assumption, some frog must occupy sector n and then goes to sector 1 to reduce that sum. Thus, the claim is justified. Consider two adjacent sectors k and $k + 1$. Suppose one of them, say sector k, is occupied by a frog. This frog either remains or is joined by another, causing one of them to go to sector $k + 1$. This frog either remains or is joined by another, causing one of them to go back to sector k. It follows that if either of two adjacent sectors has been occupied, we claim that one or the other will always be occupied. If at most half of the sectors are occupied, the occupied sectors must alternate with the unoccupied ones. The next move will bring the number of occupied sectors to more than half of n. This justifies the claim.

Solution to Problem 2.11. Fyodor collects coins. No coin in his collection is more than 10 cm in diameter. He keeps all the coins arranged side by side in a rectangular box of size 30 cm × 70 cm. He gets a new coin of diameter 25 cm. Prove that he can fit all of his coins in a square box of side length 55 cm.

Answer. Divide the 30 × 70 box into a 30 × 55 part and a 30 × 25 part, with an overlap of width 10, as shown in the left-hand figure. Divide the 55 × 55 box into a 25 × 25 part, a 30 × 55 part and a 30 × 25 part, as shown in the figure on the right. The new coin is put into the 25 × 25 part. Since no old coin has a diameter exceeding 10, each of them lies entirely within one of the two parts of the original box, and can be assigned to it. If we transfer all the coins from each part of the original box in exactly the same formation into the corresponding parts of the new box, everything will fit.

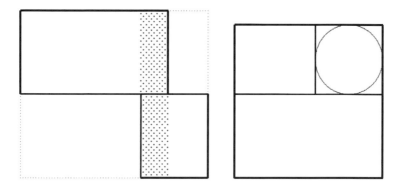

Solution to Problem 2.12. A digital clock shows the time as 21:06. How many minutes will it take until the clock shows the same four digits (in some different order) the next time?

Answer. The first time the same four digits will be displayed is at 01:26. This is 260 minutes after 21:06.

Solution to Problem 2.13. A functioning calculator shows digits in the following way:

A broken calculator shows only horizontal traces. Determine the sum of all the digits in the expression, as shown by the broken calculator in the figure.

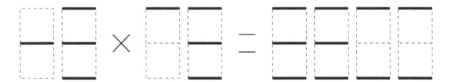

Answer. The last two digits must be 00, since this is the only digit for which the horizontal lines at the top and the bottom appear without the horizontal line in the middle. Similarly, the first digit must be 4, and the third digit must be 7, since these are the only digits with a unique horizontal line in the middle and at the top, respectively. Since the product ends in 00, it must be divisible by 100. Since the fourth digit cannot be 0, the second 2-digit number must be 75, and the first an even number divisible by 4. We can easily check $48 \times 75 = 3600$. The sum of the digits is therefore 33.

Solution to Problem 2.14. In a triangle ABC, we are given $AC = BC$, $\angle ABM = 60°$, $\angle ACB = 20°$ and $\angle BAN = 50°$. Determine the size of $\angle$ AMN.

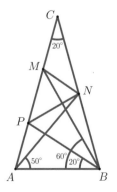

Answer. Since the triangle ABC is isosceles, we have

$$\angle ABC = \angle BAC = 80°.$$

Let P be the point on AC such that $\angle ABP = 20°$. Then $\angle PBN = 60°$. Since

$$\angle BAP = \angle PAB = 80° \quad \text{and} \quad \angle BAN = \angle BNA = 50°,$$

triangles ABP and ABN are isosceles and $AB = BP = BN$ holds. We see that BPN is equilateral, and this means that $BP = BN = PN$ holds. Let us now take a look at the triangle BPM. Since

$$\angle PBM = \angle ABM - \angle ABP = 40° \text{ and } \angle BPM = 180° - \angle APB = 100°,$$

we also have $\angle PMB = 40°$. This means that $PM = PN$ also holds. Triangle PMN is therefore isosceles with $\angle MPN = \angle MPB - \angle NPB = 100° - 60° = 40°$, and we therefore obtain $\angle PMN = 70°$.

Solution to Problem 2.15. Justin's sock drawer contains only identical black socks and identical white socks, a total of less than 50 socks altogether. If he withdraws two socks at random, the probability that he gets a pair of the same color is 0.5. What is the largest number of black socks he can have in his drawer?

Answer. If we let b and w denote the number of black socks and white socks, respectively, the probability tree gives us

$$b \cdot (b - 1) + w \cdot (w - 1) = 2 \cdot b \cdot w.$$

Rearranging, we see that this is equivalent to $b + w = (b - w)^2$. This is a property possessed only by two adjacent triangular numbers. The largest possible numbers are therefore $b = 28$ and $w = 21$.

Solution to Problem 2.16. X is a point inside an equilateral triangle ABC. Y is the foot of the perpendicular from X to AC, Z is the foot of the perpendicular from X to AB, and W is the foot of the perpendicular from X to BC. The ratio of the distances of X from the three sides of the triangle is $1:2:4$ as shown in the diagram. If the area of $AZXY$ is $13\,\text{cm}^2$, find the area of ABC.

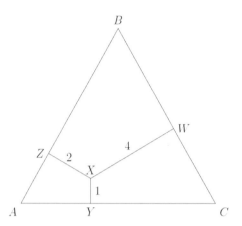

Answer. Draw a 7-layer grid of equilateral triangles each of height 1, starting with a single triangle in the top layer, then a trapezium of three triangles in the next layer, a trapezium of five triangles in the next layer, and so on. The boundary of the combined figure is $\triangle ABC$ and X is one of the grid vertices as shown. There are 49 small triangles in ABC and 6.5 small triangles in $AZXY$. Hence, after rescaling so that the area of $AZXY$ is $13\,\text{cm}^2$, the area of ABC is $13\,\text{cm} \times 49/6.5\,\text{cm} = 98\,\text{cm}^2$.

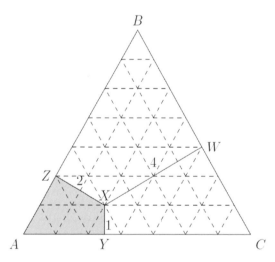

Solution to Problem 2.17. Nine stars ******** are written on a blackboard. John and Mary alternate turns, replacing them with the digits 1,

2, 3, 4, 5, 6, 7, 8, and 9. John, who starts the game, can replace any one of the stars with any one of the digits. When it is Mary's turn, she replaces any two of the remaining stars with any two of the remaining digits, and so on. After three moves, there is a 9-digit number on the blackboard, formed with each of the digits used exactly once. Mary wins if this resulting 9-digit number is divisible by 27. If it is not, John wins. Can Mary always win the game, independent of John's moves? What strategy must she apply?

Answer. Yes, Mary can always win the game. A possible winning strategy for her can be derived from the following two facts.

Fact 1: If the 3-digit number ABC is divisible by 27, then the numbers BCA and CAB are also divisible by 27. This follows from the fact that

$$10 \cdot ABC - BCA = 10 \cdot (100A + 10B + C) - (100B + 10C + A)$$
$$= 999 \cdot A.$$

Since the numbers $10 \cdot ABC$ and $999 \cdot A = 37 \cdot 27 \cdot A$ are both divisible by 27, this must also be true of number BCA. This implies that any cyclic permutation of the digits of any 3-digit number divisible by 27 yields a 3-digit number also divisible by 27.

Fact 2: The digits 1, 2, 3, 4, 5, 6, 7, 8 and 9, can be used to form three 3-digit integers, namely 972, 486 and 135, all of which are divisible by 27. From fact 1, we know that cyclic permutations of their digits yield numbers that are divisible by 27 as well.

It is now quite clear how Mary might proceed. If John replaces any of the stars with any of the digits in any spot, Mary can rotate the remaining two digits of the 3-digit number from her list, containing the digit John has chosen, in order to get an appropriate 3-digit number divisible by 27 and thus replace the two remaining stars in that third of the line of stars containing the star which John has replaced. It is clear that the 9-digit number that John and Mary compose is then certainly divisible by 27.

Solution to Problem 2.18. Thirteen bees, one little bee and twelve large bees, are living in a 37-cell honeycomb. Each large bee occupies three pairwise adjacent cells and the little bee occupies exactly one cell, as in the picture. In how many ways can the honeycomb be divided into thirteen non-overlapping sectors so that all thirteen bees can be accommodated in accordance with the given restrictions?

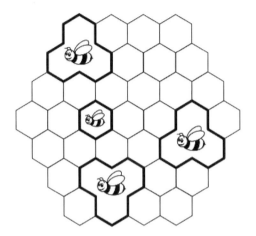

Answer. Let us consider the thirteen cells shaded as shown.

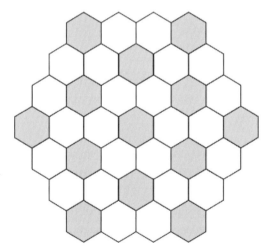

Each three-cell sector contains exactly one shaded cell, so the cell of the little bee must be one of the shaded ones. If it is the central cell, there are exactly two ways to divide the rest of the honeycomb into twelve large bee sectors (the one shown in the picture and the one rotated by 60 degrees). For each of the six 'middle' shaded cells there is exactly one way to accommodate large bees in the rest of the honeycomb. Finally, for each of the shaded cells on the boundary, there are exactly two ways to divide the remaining cells into three-cell sectors (the one shown and the symmetric one).

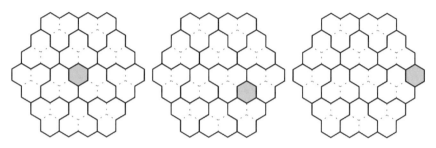

This gives a total of $2 + 6 \cdot 1 + 6 \cdot 2 = 20$ ways to dissect the honeycomb into the sectors as required.

Solution to Problem 2.19. We are given a quadrilateral $ABCD$ with side lengths as shown in the picture.

How long is $CF + FD$ if we know that $BE + EC = 78$ holds?

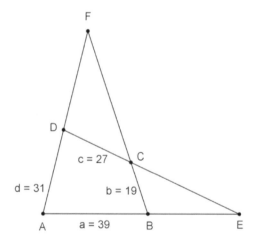

Answer. First, we show that the triangles BEC, AED, DCF, and ABF have an identical excircle lying in the interior of the angle $\angle EAF$.

Let G denote the tangent point of the B-excircle of triangle BEC at line BE. The distance between B and G is half the perimeter of the triangle,

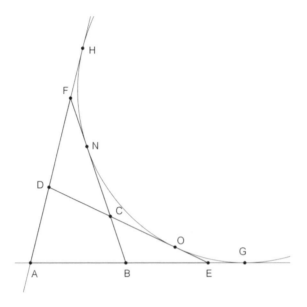

i.e. $BG = \frac{1}{2}(CB + BE + EC)$. Now let G' be the tangent point of the A-excircle of triangle AED at line AE.

Using $a + b = 58 = c + d$ we get

$$AG' = \frac{1}{2}(CE + ED + DA) = \frac{1}{2}(AB + BE + EC + CD + DA)$$

$$= \frac{1}{2}(AB + BE + EC + AB + BC) = AB + \frac{1}{2}(BE + EC + BC),$$

and hence $G = G'$. Therefore, the tangent point O on side CE is uniquely determined for both triangles, and it follows that the triangles BEC and AED have an identical excircle. Similarly, we can also show that the triangles DCF and ABF have the same excircle. It is even identical for all four triangles under consideration since the midpoint of the excircle must lie on the bisector of $\angle EAF$ and on the bisector of $\angle ECF$. Since the tangents to a circle from a common point have equal lengths, $AG = AH$ implies that the triangles AED and ABF have equal perimeters. We therefore obtain

$$CF + FD = AB + BF + FA - AB - BC - DA$$

$$= AE + ED + DA - AB - BC - DA$$

$$= BE + EC + CD - BC$$
$$= 78 + 27 - 19 = 86$$

as the length of $CF + FD$.

Solution to Problem 2.20. Gilbert Bates, a very rich man, wants to have a new swimming pool installed in his garden. Since he likes symmetry, he tells the gardener to build an elliptical pool within a $10\,\text{m} \times 10\,\text{m}$ square $ABCD$. This elliptical pool is to touch all four sides of the square, and particularly the side AB at a point P which is $2.5\,\text{m}$ from A. The gardener, who knows how to construct an ellipse if the foci and a point on the ellipse are given, realizes that, due to symmetry, he only needs to determine the distance between the foci in order to construct the ellipse. Can you help him by computing the distance between the foci in meters?

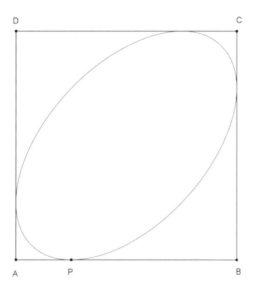

Answer. We solve the problem in a more general way. Let $ABCD$ be a square of side length 1 with corner A at the origin of a coordinate system. Point P is on AB and has coordinates $(b, 0)$ with $0 < b < \frac{1}{2}$. If the focus F_1 has coordinates (f, f), then the focus F_2 has coordinates $(1 - f, 1 - f)$, because both foci lie symmetrically on the diagonal AC with respect to the other diagonal.

Now, the line g_1 through P and F_1 is given by the equation

$$y = (x - b) \cdot \frac{f}{f - b},$$

and the line g_2 through P and F_2 by

$$y = (x - b) \cdot \frac{f - 1}{b + f - 1}.$$

Since the line through P perpendicular to the tangent AB bisects the angle $\angle F_2 P F_1$, the slope of g_2 must be the negative of the slope of g_1. Therefore, we get

$$\frac{f}{f - b} = (-1) \cdot \frac{f - 1}{b + f - 1},$$

which can be simplified to $f^2 - f + \frac{b}{2} = 0$.

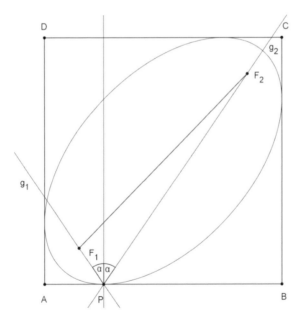

This quadratic equation has the two solutions $f_{1,2} = \frac{1}{2}(1 \pm \sqrt{1 - 2b})$, each one corresponding to one focus of the ellipse. The distance between the foci can now be computed by the Pythagorean theorem, yielding $\sqrt{2 - 4b}$. Setting $b = \frac{1}{4}$ and scaling up by 10 gives the desired result $F_1 F_2 = 10$.

Remark. The solution shows a way to construct an ellipse touching a square at a given point with gardener's ellipse and ruler and compass only.

Solution to Problem 2.21. Forty-one rooks are placed on a 10 × 10-chessboard. Prove that you can certainly choose five of them that do not attack each other. (We say that two rooks "attack" each other if they are in the same row or column of the chessboard.)

Answer. The following is just one solution out of three known to me. It was first found during the 1984 Colorado Mathematical Olympiad by Russel Shaffer. The idea of using symmetry of all colors by gluing a cylinder belongs to my undergraduate university student Bob Wood.

Let us make a cylinder out of the chessboard by gluing together two opposite sides of the board, and color the cylinder diagonally in 10 colors, as shown in the figure. Now we have $41 = 4 \times 10 + 1$ pigeons (rooks) in 10 pigeonholes (one-color diagonals), and therefore, by the Pigeonhole Principle, there is at least one hole that contains at least 5 pigeons. Since the 5 rooks located on the same one-color diagonal do not attack each other, the proof is complete.

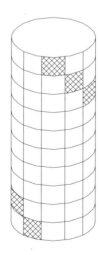

Solution to Problem 2.22. (a) Each member of two 7-member chess teams is to play once against each member of the opposing team. Prove that as soon as 22 games have been played, we can choose 4 players and seat them at a round table so that each pair of neighbors has already played.

(b) Prove that 22 is the best possible value, i.e., the result cannot be guaranteed after 21 games.

Answer. This problem occurred to me while on the shore of the Kochelsee in Bavaria. I was reading a wonderful unpublished 1989 manuscript of the monograph *Aspects of Ramsey Theory* by Hans Jürgen Prömel and Bernd Voigt. I found a mistake in a lemma and constructed a counterexample for this lemma's statement. The counterexample is Problem 22(b) as stated here. Problem 22(a) is a corrected particular case of that lemma, translated, of course, into the language of a nice "real" story. The following is one of three wonderful solutions that I found.

In the selection and editing process, Dr. Col. Bob Ewell suggested using a 7×7-table to record the games played. We number the players in both teams. For each player of the first team we allocate a row of the table and for each player of the second team a column. We place a checker in the table in location (i, j) if the player i of the first team played the player j of the second team.

If the required four players were found, this would manifest itself in the table as a rectangle formed by four checkers (which we will call a *checkered rectangle*). The problem thus translates into the new language as follows: *A 7×7-table with 22 checkers must contain a checkered rectangle.*

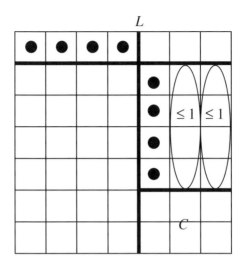

Assume that a table has 22 checkers but does not contain a checkered rectangle. Since 22 checkers are contained in 7 rows, by the Pigeonhole Principle, there is a row with at least 4 checkers in it. Observe that interchanging rows or columns does not affect the property of the table to have or not have a checkered rectangle. By interchanging rows, we make the row with at least 4 checkers first. By interchanging columns, we make all checkers appear consecutively from the left of the first column. We consider two cases.

First, we consider the case in which there are exactly four checkers in the first row. Draw a bold vertical line L after the first 4 columns. To the left from L, the top row contains 4 checkers, and all other rows contain at most one checker each, for otherwise we would have a checkered rectangle (that includes the top row). Therefore, to the left from L we have at most $4 + 6 = 10$ checkers. This leaves at least 12 checkers to the right of L, and thus at least one of the three columns to the right of L contains at least 4 checkers. By interchanging columns and rows, we put them in the position shown in the figure. Then, each of the two right columns contains at most one checker total in rows 2 through 5, for otherwise we would have a checkered rectangle. We thus have at most $4 + 1 + 1 = 6$ checkers to the right of L in rows 2 through 5 combined. Therefore, in the lower right 2×3 part C of the table we have at least $22 - 10 - 6 = 6$ checkers, which

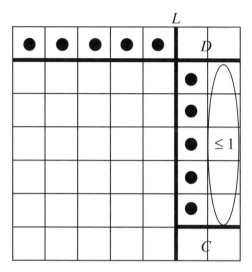

implies that C is completely filled with checkers, and we get a checkered rectangle in C in contradiction to our assumption.

We now assume that the top row contains at least 5 checkers.

Once again, we draw a bold vertical line L, this time after the first five columns. To the left from L, the top row contains five checkers, and all other rows contain at most one checker each, for otherwise we would have a checkered rectangle (that includes the top row). Therefore, to the left from L we have at most $5 + 6 = 11$ checkers. This leaves at least 11 checkers to the right of L, and thus at least one of the two columns to the right of L contains at least 6 checkers. By interchanging columns and rows, we put 5 of these 6 checkers in the position shown in the figure. The last column then contains at most one checker in total in the rows 2 through 6, for otherwise we would have a checkered rectangle. We thus have at most $5 + 1 = 6$ checkers to the right of L in rows 2 through 6 combined. Therefore, the upper right 1×2 part D of the table plus the lower right 1×2 part C of the table together have at least $22 - 11 - 6 = 5$ checkers. Since they only have 4 cells, we have a contradiction.

(b) Glue a cylinder out of the board 7×7, and put 21 checkers on all squares of the first, second, and fourth diagonals (the left-hand figure below shows the cylinder with one checkered diagonal and the right-hand figure shows the cylinder with all three cylinder diagonals in a flat representation).

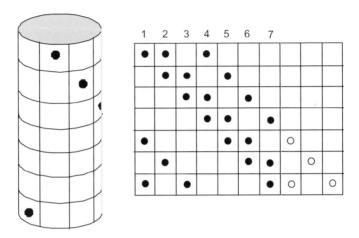

Assume that four checkers form a rectangle on our 7×7-board. Since these four checkers lie on three diagonals, the Pigeonhole Principle implies that two checkers lie on the same (checkers-covered) diagonal D of the cylinder. This means that, *on the cylinder*, our four checkers form a square! Two other (opposite) checkers a and b must therefore be symmetric to each other with respect to D, which implies that the diagonals of the cylinder that contain a and b must be symmetric with respect to D, but no two checker-covered diagonals in our checker placement are symmetric with respect to D. (To see this, consider the circular figure below, which shows the top rim of the cylinder with bold dots for checkered diagonals: square distances between the checkered diagonals, clockwise, are 1, 2, and 4.) This contradiction implies that there are no checkered rectangles in our placement, and we are done.

Remark on Problem (b). Obviously, any solution of problem (b) can be presented in the form of 21 checkers on a 7×7-board. It is less obvious that the solution is *unique*: by a series of interchanges of rows and columns, any solution of this problem can be brought to precisely the one I presented! Of course, such interchanges mean merely renumbering players of the same team. The uniqueness of the solution of problem (b) is precisely another way of stating the uniqueness of the projective plane[1] of order 2, the

---

[1] A finite projective plane of order n is defined as a set of $n^2 + n + 1$ points with the properties that:

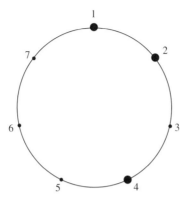

so-called "Fano Plane"[2] denoted by $PG(2,2)$. The Fano plane is an abstract construction, with symmetry between points and lines: it has 7 points and 7 lines (think of rows and columns of our 7×7-board as lines and points, respectively!), with 3 points on every line and 3 lines through every point, as illustrated.

Observe that replacing checkers by ones and the rest of the squares by zeroes in our 7×7-board yields the incidence matrix of the Fano Plane.

Solution to Problem 2.23 — Intriguing Fifties. Let S be a 5-element set. How many functions $f : S \to S$ are there, such that $f^{50}(x) = x$ for each $x \in S$? Here $f^{50} = f \circ \cdots \circ f$ (50th iterate).

Answer. Clearly f must be bijective, hence a permutation of S, and hence it splits into cycles. The condition $f^{50} = id_S$ is equivalent to requiring that all cycle lengths should be divisors of 50 (not exceeding 5, of course). In the table below, the columns (from left to right) represent:

- pattern (splitting S into subsets of cardinalities 5, 2, 1);
- number of ways to split S, in accordance to the pattern;
- number of ways to arrange the subsets into cycles;
- product of the entries in the second and third column:

---

(1) Any two points determine a line.

(2) Any two lines determine a point.

(3) Every point has $n + 1$ lines through it.

(4) Every line contains $n + 1$ points.

[2]Named after Gino Fano (1871–1952), the Italian geometer who pioneered the study of finite geometries.

5		1	4! 24
$2+2+1$	$\binom{5}{2}\binom{3}{2}\cdot\frac{1}{2}$		1 15
$2+1+1+1$	$\binom{5}{2}$		1 10
$1+1+1+1+1$	1		1 1

The sum of the numbers in the last column gives a total of 50, which is the answer to the question.

Some comments. Many years ago, based on the same idea, I also proposed a similar problem for the very enjoyable Canadian journal *Crux Mathematicorum*, which became Problem 2080 (they apply consecutive numbering, over years and decades):

Let T be a 7-element set. How many functions $f : T \to T$ are there, such that $f^{2080} = id_T$? Again, the answer was 2080.

One might consider an analogous question with an m-element set M and mapping $f : M \to M$ with $f^n = id_M$. Surprisingly, it is not at all rare that the number of such mappings happens to be exactly n. Obviously, this is always the case for $n = 1$ and $n = m!$, but that's no fun; interesting are instances where $1 < n < m!$. For $m = 5$, the following n (between 1 and 5!) are good: 21, 25, 26, 45, 50, 56, 66, 80, 90, 96.

Apart from numerical experimentation, I haven't heard of any theory behind that.

Solution to Problem 2.24 — Exceed Galaxy Size. Imagine an $n \times n \times n$ cube as a loose structure composed of n^3 unit cubes. The task is to split it into n^3 single cubes. This of course can be done with one efficient kick. Ignoring this kind of triviality, we introduce a rule to be observed: at any one time, it is allowed to remove any single small cube, whose upper face is visible from above. The decomposition task can be completed in many ways (really many; such combinatorial quantities like to grow rapidly...).

So, what is the smallest n for which the number of ways exceeds the diameter of our galaxy, measured in kilometers?

Answer. At each step, we have a choice from among n^2 options, the columns of the big cube. Label those columns $\alpha, \beta, \gamma, \ldots$, characters of an alphabet of cardinality n^2. The decomposition procedure corresponds to a word over this alphabet (like $\gamma\pi\pi\alpha\lambda\alpha\ldots$, for example) of length n^3, in which every character appears exactly n times. There are $\binom{n^3}{n}$ ways to locate the positions of character α; once these are located, we are left with $\binom{n^3-n}{n}$ possible positions for β, and so on. In all, there are N ways leading to complete splitting, where

$$N = \binom{n^3}{n}\binom{n^3-n}{n}\binom{n^3-2n}{n}\cdots\binom{2n}{n}\binom{n}{n} = \frac{(n^3)!}{(n!)^{n^2}}.$$

A light year is slightly less than 10^{13} km. Therefore, we need $N > 3 \cdot 10^{18}$. For $n = 2$, we obtain $N = 2520$, which is not enough. For $n = 3$, calculating N is a job for a standard (?) calculator. Why not try just working out an estimate (from above, from below?). For instance, using the inequality $n! > (\frac{n}{3})^n$ (easily proved by induction) for $n = 3$, we obtain

$$N = \frac{27!}{6^9} > \frac{9^{27}}{6^9} = \left(\frac{9^3}{6}\right)^9 = \left(\frac{243}{2}\right)^9 > 1.2^9 \cdot 10^{18} > 5 \cdot 10^{18},$$

and this seems to be quite enough. (Actually, $N = 1.080491954750208 \cdot 10^{21}$).

So, $n = 3$ is the answer. (Was that your guess?)

Solution to Problem 2.25 — A Swelling Ball. In a square with side-length 2, you can place four circles of diameter 1 in the corners, with each of them touching two sides of the square and two neighboring circles. In the center there is some space left for a tiny circle, touching those four. Similarly, in a cube with edge-length 2, one can accommodate eight balls of diameter 1 in the corners, with each touching three faces and three other balls; and a small ball, concentric with the cube, tangent to the eight diameter-1 balls. The same can be done in an n-dimensional cube of edge-length 2, with 2^n diameter-1 n-dimensional balls (hyperspheres) in the corners and another ball, of radius r_n, fitting in the central region and tangent to those 2^n ones. It is tempting to conjecture that $r_n \to 1$ as $n \to \infty$. Prove or disprove this.

Answer. It is quite easy to disprove this by simple calculation, with no tricks. The centers of the diameter-1 balls form an n-cube with edge-length 1. Let A, B be opposite centers of two balls, such that AB is a maximal diagonal, of length $\sqrt{n}$. The spherical surfaces of those balls intercept a segment of length $\sqrt{n} - 1$ on AB (because each of those balls has radius $\frac{1}{2}$). This segment is the diameter of the "centric" ball, which means that $r_n = \frac{\sqrt{n}-1}{2} \to \infty$ as $n \to \infty$.

(In dimension $n = 9$, the "centric" ball already has a radius of $r_9 = 1$ and touches all eighteen faces of the cube, after which it keeps on swelling in dimension and size ...)

Solution to Problem 2.26 — A Rapidly Increasing Sequence. Consider the rapidly growing sequence $a_n = \frac{n!}{n+1}$ and define $c_n = \prod_{k=0}^{n} a_k^{2^{n-k}}$ (for $n = 0, 1, 2, \ldots$). How many terms of the sequence $c_0, c_1, c_2, \ldots$ are numbers not greater than 2018?

Answer. All of them. Clearly, $c_0 = a_0 = 1$ and $c_{n+1} = c_n^2 \cdot a_{n+1}$. Hence, by simple induction we have $c_n = \frac{1}{(n+1)!}$.

Solution to Problem 2.27. A positive integer n is called *fantastic* if there exist positive rational numbers a and b such that $n = a + \frac{1}{a} + b + \frac{1}{b}$.

(a) Prove that there exist infinitely many prime numbers p such that no multiple of p is fantastic.
(b) Prove that there exist infinitely many prime numbers p such that some multiple of p is fantastic.

Answer. First, we note that we can write

$$r(a, b) := a + \frac{1}{a} + b + \frac{1}{b} = \frac{(a+b)(ab+1)}{ab}.$$

We put $a = \frac{t}{u}$ and $b = \frac{v}{w}$, where t, u, v and w are positive integers such that t and u are coprime, as are v and w. This gives us $r(a, b) = \frac{(tv+uw)(tw+uv)}{tuvw}$, and we see that we need to investigate the Diophantine equation

$$tu(v^2 + w^2) + vw(t^2 + u^2) = kptuvw. \tag{$*$}$$

We note that $\gcd(tu, t^2 + u^2) = 1$, and $(*)$ therefore implies $tu|vw$. As we similarly get $vw|tu$ as well, we obtain $tu = vw$, and $(*)$ therefore becomes

$$\frac{(v^2 + t^2)(v^2 + u^2)}{v^2} = t^2 + u^2 + v^2 + w^2 = kptu.$$

We see that p must either divide $v^2 + t^2$ or $v^2 + u^2$. In the case $p \equiv -1 \pmod 4$, (i.e. when -1 is a quadratic non-residue mod p), this means that p divides v (and t or u). But since the same argument is valid for w instead of v, we have $p|v$, w contradicting the coprimality of v and w. Thus, the infinitely many primes with $p \equiv -1 \pmod 4$ have no fantastic multiple and part (a) is solved.

For part (b) we choose $v = 1$ and substitute $w = tu$. Thus, we are now looking for integers t and u such that

$$1 + t^2 + u^2 + t^2 u^2 = kptu.$$

Here we choose[3] $t = F_{2l+1}$, $u = F_{2l-1}$ and use the identity[4] $1 + F_{2l+1}^2 = F_{2l+3}F_{2l-1}$ to obtain $(1 + t^2)(1 + u^2) = (1 + F_{2l+1}^2)(1 + F_{2l-1}^2) = F_{2l+3} \cdot F_{2l-1} \cdot F_{2l+1} \cdot F_{2l-3} = kptu = kp \cdot F_{2l+1} \cdot F_{2l-1}$, i.e. $F_{2l+3} \cdot F_{2l-3} = kp$. Therefore, every prime factor of the Fibonacci number F_{2l+3} has a fantastic multiple.

In view of the well-known formula $\gcd(F_a, F_b) = F_{\gcd(a, b)}$, it is clear that F_a and F_b are relatively prime if a and b are different prime numbers. Hence, we know that infinitely many prime numbers have a fantastic multiple, which solves part (b).

Open Questions

(1) What about the primes, not covered in the proofs of (a) and (b)?
(2) Of course, all fantastic numbers are at least 4.
    Let $a(n)$ be the number of all fantastic numbers in $\{4, 5, \ldots, n\}$, $n \geq 4$. Determine, in case of existence, the natural density $d(F)$ of all fantastic numbers, that is $d(F) = \lim_{n \to \infty} \frac{a_n}{n}$. In case $d(F)$ does not exist, determine the lower and upper natural densities of all fantastic numbers.
(3) Extend the concept of fantastic numbers to k-fantastic numbers, $k \geq 3$, that are natural numbers n of the type $n = a_1 + \frac{1}{a_1} + \ldots + a_k + \frac{1}{a_k}$, where $a_1, \ldots, a_k$ are suitable positive rational numbers.
    Investigate interesting properties of these numbers and develop a "sound" theory about them.

---

[3] It is a well-known problem that $tu|t^2 + u^2 + 1$ with $t > u$ is only possible if t and u are Fibonacci numbers of the form $t = F_{2l+1}$, $u = F_{2l-1}$ in which case $t^2 + u^2 + 1 = 3tu$.
[4] This is a special case of Vajda's identity $F_{n+i} \cdot F_{n+j} - F_n \cdot F_{n+i+j} = (-1)^n \cdot F_i \cdot F_j$.

Solutions to Problem 2.28. Three circles of equal radii are situated inside square *ABCD*. The first circle is tangent to sides *BC* and *CD*. The second circle, disjoint from the first, touches sides *CD* and *DA*. The third circle touches the second one and the side *DA*. The line through *A*, tangent to the third circle, and the line through *B*, tangent to the first circle, meet at a point *E* on side *CD*. Prove that the inradius of triangle *ABE* is twice as large as the radius of the three former circles.

Solution 1 by Marcin E. Kuczma. In his charming talk on the Art of Sangaku, Hidetoshi Fukagawa mentioned this problem as waiting for solution. As can be seen, my proposed proof involves quite a great deal of calculation; a nice geometric solution is still waiting to be discovered.

We take the common radius of the three circles as our unit and begin with some notation. Let *FG* be the common tangent line of the two circles (the second and the third), with *F* on *AE*, and *G* on *AD*, and let

$$AB = BC = CD = DA = a, CE = u, DE = w, GF = v, AG = x$$

(so $a - x = DG = 2$). Then (with $[XYZ]$ denoting the area of triangle XYZ), we have $au = 2 \cdot [BCE] = a + u + \sqrt{a^2 + u^2}$ (perimeter times inradius, which is equal to 1), which gives us $\sqrt{a^2 + u^2} = au - a - u$. A similar calculation with regard to $2 \cdot [AGF]$ gives us $\sqrt{x^2 + v^2} = xv - x - v$.

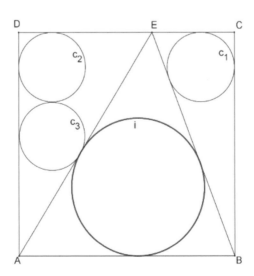

Solving the first of these equations for u, setting $a = x + 2$, and solving the second equation for v then yields

$$u = \frac{2(x+1)}{x} \quad \text{and} \quad v = \frac{2(x-1)}{x-2}.$$

From the similar triangles ADE and AGF, we have $\frac{w}{v} = \frac{a}{x} = \frac{AE}{AF}$, and therefore

$$w = \frac{av}{x} = \frac{(x+2)v}{x}.$$

Substitution in the obvious equality $u + w = a$, therefore gives us

$$\frac{2(x+1)}{x} + \frac{x+2}{x} \cdot \frac{2(x-1)}{x-2} = x+2,$$

which simplifies to $x^3 - 4x^2 - 4x + 8 = 0$.

Let r be the inradius of triangle ABE. Again, by the perimetric formula, we have

$$r \cdot (AB + AE + BE) = 2 \cdot [ABE] = a^2 = (x+2)^2.$$

To obtain the desired result $r = 2$, it will be enough to show that

$$AB + AE + BE = \frac{(x+2)^2}{2}$$

holds. Indeed, applying the results we have established gives us

$$AB + AE + BE = a + \frac{a}{x} \cdot AF + BE = a + \frac{a}{x} \cdot \sqrt{x^2 + v^2} + \sqrt{a^2 + u^2}$$

$$= a + \frac{a}{x} \cdot (xv - x - v) + (au - a - u)$$

$$= (a-1) \cdot u + \frac{a \cdot (x-1)}{x} \cdot v - a.$$

It remains to set $a = x + 2$ and plug in the values for u and v, which yields

$$AB + AE + BE = (x+1) \cdot \frac{2(x+1)}{x} + \frac{(x+2)(x-1)}{x}$$

$$\cdot \frac{2(x-1)}{x-2} - (x+2) = \frac{3x^2 - 8}{x-2},$$

and to show that these two values are equal, i.e.

$$\frac{3x^2 - 8}{x - 2} = \frac{(x + 2)^2}{2}.$$

Upon cross multiplication, this is exactly the cubic result we have already established. The proof is therefore complete.

Solution 2 by Walther Janous. (Dedicated to the memory of Gilbert Helmberg (1928–2019)). We wish to prove that $R = 2r$, where R is the inradius of the triangle ABE and r is the common inradius of the three small circles.

Without loss of generality, we let $ABCD$ be the unit-square. Furthermore, let t denote the length of CE.

Then t satisfies $0 < t < 1$ and we have $BE = \sqrt{t^2 + 1}$ and $AE = \sqrt{t^2 - 2t + 2}$.

As c_1 is the incircle of the right triangle BCE, its radius is equal to $r = \frac{2 \cdot [BCE]}{BC + CE + BE}$, that is

$$r = \frac{t}{1 + t + \sqrt{t^2 + 1}}.$$

(As before, $[XYZ]$ stands for the area of triangle XYZ.)

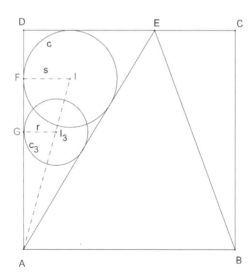

Now, we come to the crucial step of getting a reasonable expression for the radii of c_2 and c_3.

Clearly, c_3 is a smaller copy of the incircle c of the right-angled triangle AED with inradius

$$s = \frac{2 \cdot [AED]}{AE + DE + AD},$$

that is

$$s = \frac{1 - t}{2 - t + \sqrt{t^2 - 2t + 2}}.$$

The two triangles AIF and AI_3G are similar to each other. (Here I and I_3 denote the respective centers, and F and G are the two points of tangency with side AD.) Therefore, we have $\frac{AF}{s} = \frac{AG}{r}$. From $AF = 1 - s$ and $AG = 1 - 3r$, we get $\frac{1-s}{s} = \frac{1-3r}{r}$, and therefore $r = \frac{s}{2s+1}$. Inserting these expressions yields

$$\frac{t}{t + 1 + \sqrt{t^2 + 1}} = \frac{\frac{1-t}{2-t+\sqrt{t^2-2t+2}}}{2 \cdot \frac{1-t}{2-t+\sqrt{t^2-2t+2}} + 1}$$

which is equivalent to

$$\frac{t}{t + 1 + \sqrt{t^2 + 1}} = \frac{1 - t}{\sqrt{t^2 - 2t + 2} - 3t + 4}.$$

Clearing denominators, we get

$$t \cdot (\sqrt{t^2 - 2t + 2} - 3t + 4) = (1 - t)(t + 1 + \sqrt{t^2 + 1}),$$

which is equivalent to

$$t \cdot \sqrt{2 - 2t + t^2} - (1 - t) \cdot \sqrt{t^2 + 1} = 2t^2 - 4t + 1.$$

Let $l(t)$ and $r(t)$ be the expressions on the left and right side of this equation, respectively. For $0 < t < 1$, the function $l(t)$ increases, while $r(t)$ decreases. This fact, $l(0) < r(0)$, and $l(1) > r(1)$ guarantee the existence of a unique solution $t_0 \in [0; 1]$ of the equation. Furthermore, squaring the equation and simplifying gives us the equivalent equation

$$8t^3 - 28t^2 + 28t - 7 = 0,$$

and it is an easy exercise to verify that this cubic equation has a unique solution in the interval $[0; 1]$.

In order to verify the claim $R = 2r$ we now note

$$R = \frac{2 \cdot [ABE]}{AB + BE + AE}$$

which is equivalent to

$$R = \frac{1}{1 + \sqrt{t^2 - 2t + 2} + \sqrt{t^2 + 1}}.$$

It therefore remains to show that t_0 also satisfies

$$\frac{1}{1 + \sqrt{t^2 - 2t + 2} + \sqrt{t^2 + 1}} = \frac{2t}{t + 1 + \sqrt{t^2 + 1}},$$

and indeed, squaring, simplifying and proceeding as before also yields the equation

$$8t^3 - 28t^2 + 28t - 7 = 0,$$

and the proof is complete.

Remark. Solving this cubic equation by classical methods yields $t_0 = \frac{7}{6} - \frac{\sqrt{7} \cdot \sin\left(\frac{\arctan \frac{\sqrt{3}}{9}}{3} + \frac{\pi}{3}\right)}{3}$ satisfying $0 < t < 1$, i.e. $t_0 = 0.3765101981\ldots$.

Solution to Problem 2.29. Prove that there exist continuous functions $a_1, a_2, a_3, \ldots : [0, 1] \to (0, +\infty)$ such that

(i) $\sum_{n=1}^{\infty} a_n(t) < +\infty$, for all $t \in [0, 1]$.
(ii) For every sequence (b_n) of positive real numbers with $\sum_{n=1}^{\infty} b_n < +\infty$ there exists a $t \in [0, 1]$, such that $\lim_{n\to\infty} \frac{b_n}{a_n(t)} = 0$.

Answer. Let us define initially the functions a_k in the usual ternary Cantor set $K = \left\{ \sum_{m=1}^{\infty} \frac{\sigma_m}{3^m}, \sigma_m \in \{0, 2\}, \forall m \right\}$, and then extend them linearly in the connected components of the complement of K.

Given $t = \sum_{m=1}^{\infty} \frac{\sigma_m}{3^m} \in K$, we define, for $1 \leq k < +\infty$,

$$c_k(t) = \sum_{j=1}^{\infty} \frac{\sigma_{2^{k-1}(2j-1)}}{2^{j+1}}.$$

These functions are continuous and $(c_1(t), c_2(t), \ldots)$ defines a surjection from K over $[0, 1]$.

We now define $d_1(t) = c_1(t)$, and, for each $n \in \mathbb{N}$, $d_{n+1}(t) = \min\{c_{n+1}(t),$ $1 - \sum_{k=1}^{n} c_k(t)\}$. The functions d_n are continuous, non-negative and $\sum_{n=1}^{\infty} d_n(t) \le 1$ for every $t \in [0, 1]$.

Note that, if $\sum_{k=1}^{\infty} c_k(t) \le 1$ then $d_n(t) = c_n(t)$ for every n.

Finally, we define $a_n(t) = \frac{1}{2^n} + d_n(t)$. We have $a_n(t) > 0$, for all $n \in \mathbb{N}$ and all $t \in [0, 1]$, a_n continuous for every n and $\sum_{n=1}^{\infty} a_n(t) \le 2$, for all $t \in [0, 1]$.

Let now (b_n) be a sequence of positive terms with $\sum_{n=1}^{\infty} b_n < +\infty$. For every $n \in \mathbb{N}$, let $r_n = \sum_{k=n}^{\infty} b_k$. Furthermore, let $d_n = \frac{1}{2\sqrt{r_1}} \cdot \frac{b_n}{\sqrt{r_n}}$. Since $\lim_{n \to \infty} r_n = 0$, it follows that $\lim_{n \to \infty} \frac{b_n}{d_n} = 0$.

Moreover, $d_n \le \frac{1}{\sqrt{r_1}} \cdot \frac{b_n}{\sqrt{r_n} + \sqrt{r_{n+1}}} = \frac{1}{\sqrt{r_1}} \cdot \frac{r_n - r_{n+1}}{\sqrt{r_n} + \sqrt{r_{n+1}}} = \frac{1}{\sqrt{r_1}} \left(\sqrt{r_n} - \sqrt{r_{n+1}} \right)$, thus

$$\sum_{n=1}^{\infty} d_n \le \frac{1}{\sqrt{r_1}} \sum_{n=1}^{\infty} (\sqrt{r_n} - \sqrt{r_{n+1}}) = \frac{\sqrt{r_1}}{\sqrt{r_1}} = 1.$$

So, there exists $t \in [0, 1]$ such that $d_n = d_n(t)$, for all n. Since $d_n(t) < a_n(t)$, for all $n \in \mathbb{N}$ and $\lim_{n \to \infty} \frac{b_n}{d_n(t)} = 0$, it follows that $\lim_{n \to \infty} \frac{b_n}{a_n(t)} = 0$.

Solution to Problem 2.30. Determine the smallest positive integer n for which a positive integer k exists, such that the last 2012 digits in the decimal representation of n^k are equal to 1.

Answer. This solution is due to Franco Matheus de Alencar Severo from Rio de Janeiro. To solve this problem we should find n such that there exists a value of k with

$$n^k \equiv \underbrace{11\ldots1}_{2012} = \frac{10^{2012} - 1}{9} \pmod{10^{2012}} \Leftrightarrow \begin{cases} 9n^k \equiv -1 \pmod{5^{2012}} & \text{(I)}, \\ 9n^k \equiv -1 \pmod{2^{2012}} & \text{(II)}. \end{cases}$$

Notice that every perfect square is congruent to 0, 1 or 4 modulo 8, so we cannot have k even, since we would otherwise have $9 \cdot n^k \equiv 0, 1, 4 \pmod{8}$, in contradiction to (II).

Therefore, k is odd, and it follows that $n \equiv 7 \pmod{16}(*)$ must hold, since we would otherwise also have a contradiction to (II). Indeed, since n^2 is congruent to 1 modulo 8, and n is congruent to 111, which is congruent to -1 modulo 8. This implies that n^2 is congruent to 1 modulo 16, and so n is congruent to n^k modulo 16, and n^k is congruent to 1111, which is congruent to 7 modulo 16.

The multiplicative order of n modulo 5 is a divisor of k, since the last digit of n^k is equal to 1. On the other hand, by Fermat's theorem, it should also be a divisor of 4, and it is therefore equal to 1, which yields $n \equiv 1$ (mod 5) (∗∗).

Notice that the smallest positive integer which satisfies (∗) and (∗∗) simultaneously is $n = 71$.

We will show that $n = 71$ is a solution. In order to do this, we will use the following two lemmas — we recall that, if $\gcd(a, n) = 1$, $\mathrm{ord}_n a$ denotes the multiplicative order of a modulo n:

Lemma 1. $\mathrm{ord}_{5^t} 71 = 5^{t-1}$, *for all* $t \geq 1$.

Proof. We will prove by induction on t that $5^t \parallel 71^{5^{t-1}} - 1$ holds. (We recall that, given p prime, $k \geq 2$ and $n \in \mathbb{Z}$, $p^k \parallel n$ if and only if $p^k \nmid n$ and $p^{k+1} \nmid n$ hold.) This is obviously true for $t = 1$. If it holds for some t, we have $71^{5^{t-1}} = 5^t \cdot l + 1$, where $5 \nmid l$, and therefore:

$$71^{5^t} = (5^t \cdot l)^5 + 5 \cdot (5^t \cdot l)^4 + 10 \cdot (5^t \cdot l)^3 + 10(5^t \cdot l)^2 + 5^{t+1}l + 1$$

$$\equiv 5^{t+1}l + 1 (\mathrm{mod}\ 5^{t+2}) \Rightarrow 5^{t+1} \parallel 71^{5^t} - 1.$$

From $5^{t-1} \parallel 71^{5^{t-2}} - 1$, $5^t \parallel 71^{5^{t-1}} - 1$, it follows that $\mathrm{ord}_{5^t} 71$ is a divisor of 5^{t-1}, but not a divisor of 5^{t-2}, and so we have $\mathrm{ord}_{5^t} 71 = 5^{t-1}$, which concludes the proof of the Lemma. □

Lemma 2. $\mathrm{ord}_{2^t} 71 = 2^{t-3}$, *for all* $t \geq 4$.

Proof. We will prove by induction on t that $2^t \parallel 71^{2^{t-3}} - 1$, $\forall t \geq 4$.
It is easy to check that this is true for $t = 4$. If it is true for t, then

$$71^{2^{t-3}} = 2^t \cdot l + 1, \text{ where 2 does not divide } l, \text{ thus } 71^{2^{t-2}}$$

$$= 2^{2t}l^2 + 2^{t+1}l + 1 \equiv 2^{t+1}l + 1 (\mathrm{mod}\ 2^{t+2})$$

$$\Rightarrow 2^{t+1} \parallel 71^{2^{t-2}} - 1.$$

It is easy to check that $\mathrm{ord}_{2^4} 71 = 2$. For $t \geq 5$, from $2^{t-1} \parallel 71^{2^{t-4}} - 1$, $2^t \parallel 71^{2^{t-3}} - 1$, it follows, as before, that $\mathrm{ord}_{2^t} 71 = 2^{t-3}$. □

Notice that, since $71^k \equiv 1 (\mathrm{mod}\ 5)$, the possible remainders of the division of 71^k by 5^{2012} are: $1, 6, 11, \ldots, 5^{2012} - 5 + 1$ (5^{2011} possible values). Since the order of 71 modulo 5^{2012} is 5^{2011} by Lemma 1, it follows that 71^k

assume all the mentioned possible remainders. In particular $\frac{10^{2012}-1}{9} \equiv 1$ (mod 5), so there exists a k_1 such that $71^{k_1} \equiv \frac{10^{2012}-1}{9}$ (mod 5^{2012}). Since $71 \equiv 7$ (mod 16), and $7^2 \equiv 1$(mod 16), we have $71^k \equiv 7, 1$ (mod 16), so the possible remainders are the 2^{2009} numbers

$$1, 7, 17, 23, \ldots, 2^{2012} - 16 + 1, 2^{2012} - 16 + 7 \text{ in the division by } 2^{2012}.$$

Since $\mathrm{ord}_{2^{2012}}\, 71 = 2^{2009}$, we see that 71^k assumes all the mentioned remainders, and, in particular, since $\frac{10^{2012}-1}{9} \equiv (-1) \cdot 9 \equiv 7$ (mod 16), there exists a k_2 such that $71^{k_2} \equiv \frac{10^{2012}-1}{9}$(mod 2^{2012}). Since $\gcd(2^{2009}, 5^{2011}) = 1$, the Chinese remainder theorem shows us that there exists a positive integer k with

$$\begin{cases} k \equiv k_1 \pmod{5^{2011}} \\ k \equiv k_2 \pmod{2^{2009}} \end{cases} \rightarrow 71^k \equiv \frac{10^{2012}-1}{9}\pmod{10^{2012}}.$$

Index

A

angles of a triangle, 7, 14
arbelos, 53
arithmetic mean, 111

B

basketball, 27, 30–31
beach ball, 27, 29–30, 40
beauty of mathematics, 117

C

centroid, 69
chessboard, 136, 156
circles, 47
circle of inversion, 48
circumcircle, 101, 103–105
circumscribed sphere, 33
consecutive integers, 5, 10
continuous functions, 114, 139, 170
cube, 34–35, 39, 137, 162

D

decimal representation, 89, 139, 171
discontinuous functions, 110
dodecahedron, 28, 39

E

ellipse, 135, 154
equilateral triangle, 25, 67, 74, 78, 133, 148
Euler's formula, 19, 39, 43

F

Fano plane, 161
fractions, 96
functional equation, 109
functions, 137, 161

G

generalizations, 118
golf ball, 23, 33
graph, 144

H

Heron's formula, 77
honeycomb, 135, 150

I

incenter, 105
incircle, 67, 73, 82
inequality, 9
inscribed sphere, 33
inverse points, 48
inversion, 47
irregular balls, 21

M

mathematical induction, 120
Miquel's six circle theorem, 51–52

N

number puzzles, 129

O

octahedron, 31
orthocenter, 103, 105

P

Pappus of Alexandria, 54
parallelogram, 103–104
period, 119
Pigeonhole Principle, 156, 158, 160
polyhedra, 17
prime numbers, 138
primes, 5, 10, 91
probability, 4, 9, 133, 148

R

rectangle, 7, 14, 73, 77, 97, 102–103
regular hexagons, 19, 24, 93
regular octahedron, 34–35
regular pentagons, 19, 24
rhombus, 66, 75, 78, 95
right triangle, 71–73, 80, 82, 102
roundness, 17, 33–34, 39

S

Sangaku, 61, 98, 138, 166
semicircle, 70, 99, 101

sequence, 137, 139, 164, 170
Sierpinski's triangle, 119
soccer ball, 24–25, 27
soroban, 62
sphere triangulation, 39
square, 64, 68, 79–81, 84–85, 95, 100, 123, 139, 166
Steiner's theorem, 55
Steinitz' theorem, 43
strategy, 150

T

table, 90–91
tennis ball, 26, 33
tessellations, 42
tetrahedron, 39
toroidal balls, 45
trapezoid, 65, 71, 98, 104
truncated icosahedron, 28, 37, 39

V

volleyball, 27, 31

W

wasan, 62
winning strategy, 131